Ferdinand Verhulst

# Nonlinear Differential Equations and Dynamical Systems

With 107 Figures

Springer-Verlag Berlin Heidelberg New York
London Paris Tokyo Hong Kong

Ferdinand Verhulst
Department of Mathematics, University of Utrecht
Budapestlaan 6, Postbus 80.010
NL-3508 TA Utrecht, The Netherlands

Title of the original Dutch edition
Nietlineaire Differentiaalvergelijkingen en Dynamische Systemen
Published by Epsilon Uitgaven, Utrecht 1985


Mathematics Subject Classification (1980): 34-02, 58F, 70K


ISBN 3-540-50628-4 Springer-Verlag Berlin Heidelberg New York
ISBN 0-387-50628-4 Springer-Verlag New York Berlin Heidelberg

Library of Congress Cataloging-in-Publication Data
Verhulst, F. (Ferdinand), 1939 – Nonlinear differential equations and dynamical
systems / Ferdinand Verhulst.
p. cm. -- (Universitext). Includes bibliographical references.
ISBN 0-387-50628-4 (U.S. : alk. paper)
1. Differential equations, Nonlinear. 2. Differentiable dynamical systems. I. Title.
QA372.V.47 1989   515'.355--dc20   89-21770

© Springer-Verlag Berlin Heidelberg 1990
Printed in Germany

Printing and binding: Weihert-Druck GmbH, Darmstadt
2141/3140–543210 – Printed on acid-free paper

# Universitext

# Preface

This book was written originally in Dutch for Epsilon Uitgaven; the English version contains a number of corrections and some extensions, the largest of which are sections 11.6-7 and the exercises.

In writing this book I have kept two points in mind. First I wanted to produce a text which bridges the gap between elementary courses in ordinary differential equations and the modern research literature in this field. Secondly to do justice to the theory of differential equations and dynamical systems, one should present *both* the qualitative and quantitative aspects.

Thanks are due to a number of people. A.H.P. van der Burgh read and commented upon the first version of the manuscript; also the contents of section 10.2 are mainly due to him.

Many useful remarks have been made by B. van den Broek, J.J. Duistermaat, A. Doelman, W. Eckhaus, A. van Harten, E.M. de Jager, H.E. Nusse, J.W. Reyn and a number of students. Some figures were produced by E. van der Aa, B. van den Broek and I. Hoveijn. J. Grasman and H.E. Nusse kindly consented in the use of some figures from their publications.

The typing and TEX-editing of the text was carefully done by Diana Balk.

Writing this book was a very enjoyable experience. I hope that some of this pleasure is transferred to the reader when reading the text.

Utrecht                                                                      Ferdinand Verhulst
April 1989

# Contents

# 1 Introduction

## 1.1 Definitions and notation

This section contains material which is basic to the development of the theory in the subsequent chapters. We shall consider differential equations of the form

$$\dot{x} = f(t, x) \tag{1.1}$$

using Newton's fluxie notation $\dot{x} = dx/dt$. The variable $t$ is a scalar, $t \in \mathbb{R}$, often identified with time. The vector function $f : G \to \mathbb{R}^n$ is continuous in $t$ and $x$; $G$ is an open subset of $\mathbb{R}^{n+1}$, so $x \in \mathbb{R}^n$.

The vector function $x(t)$ is a solution of equation 1.1. on an interval $I \subset \mathbb{R}$ if $x : I \to \mathbb{R}^n$ is continuously differentiable and if $x(t)$ satisfies equation (1.1).

All variables will be real and the vector functions will be real-valued unless explicitly stated otherwise. Apart from equations of the form (1.1) we shall also consider scalar equations like

$$\ddot{y} + y = y^2 + \sin t \tag{1.2}$$

in which $\ddot{y} = d^2y/dt^2$.

Equation (1.2) can be put into the form of equation (1.1) by introducing $y = y_1$, $\dot{y} = y_2$ which yields the system

$$\begin{aligned} \dot{y}_1 &= y_2 \\ \dot{y}_2 &= -y_1 + y_1^2 + \sin t \end{aligned} \tag{1.3}$$

Equation (1.3) is clearly in the form of equation (1.1); the vector $x$ in (1.1) has in the case of (1.3) the components $y_1$ and $y_2$, the vector function $f$ has the components $y_2$ and $-y_1 + y_1^2 + \sin t$. It is easy to see that the general $n^{th}$ order scalar equation

$$d^n x/dt^n = g(t, x, dx/dt, \ldots, d^{n-1}x/dt^{n-1})$$

with $g : \mathbb{R}^{n+1} \to \mathbb{R}$, can also be put into the form of equation (1.1).

Depending on the formulation of the problem, it will be useful to have at our disposal both scalar equations and their vector form.

In some parts of the theory we shall make use of the Taylor expansion of a vector function $f(t, x)$. The notation $\partial f / \partial x$ indicates the derivative with respect to the (spatial) variable $x$; this means that $\partial f / \partial x$ is the $n \times n$ matrix

$$
\begin{pmatrix}
\frac{\partial f_1}{\partial x_1} & \cdots & \frac{\partial f_1}{\partial x_n} \\
\vdots & & \vdots \\
\frac{\partial f_n}{\partial x_1} & \cdots & \frac{\partial f_n}{\partial x_n}
\end{pmatrix}
$$

with $x_1, \ldots, x_n$ the components of $x$ and $f_1, \ldots, f_n$ the components of $f(t, x)$. If we do not state an explicit assumption about the differentiability of the vector function $f(t, x)$, we assume it to have a convergent Taylor expansion in the domain considered. More often than not, this assumption can be weakened if necessary. In a number of cases we shall say that the vector function is smooth, which means that the function has continuous first derivatives.

In studying vector functions in $\mathbf{R}^n$ we shall employ the norm

$$
\|f\| = \sum_{i=1}^{n} |f_i|.
$$

For the $n \times n$ matrix $A$ with elements $a_{ij}$ we shall use the norm

$$
\|A\| = \sum_{i,j=1}^{n} |a_{ij}|.
$$

In the theory of differential equations the vector functions depend on variables like $t$ and $x$. We shall use some straightforward generalisations of concepts in basic analysis. For instance the expression

$$
\int x \, dt
$$

is short for the vector field with components

$$
\int x_i(t) \, dt \, , i = 1, \ldots, n.
$$

In the same way $\int A(t) \, dt$ represents a $n \times n$ matrix with elements $\int a_{ij}(t) \, dt \, , i, j = 1, \ldots, n$.

Finally, in estimating the magnitude of vector functions, we shall use the uniform norm or supnorm. Suppose that we are considering the vector function $f(t, x)$ for $t_0 \leq t \leq t_0 + T$ and $x \in D$ with $D$ a bounded domain in $\mathbf{R}^n$; then

$$
\|f\|_{\text{sup}} = \sup_{\substack{t_0 \leq t \leq t_0 + T \\ x \in D}} \|f\|
$$

## 1.2 Existence and uniqueness

The vector function $f(t, x)$ in the differential equation (1.1) has to satisfy certain conditions, of which the Lipschitz condition is the most important one.

**Definition**

Consider the function $f(t, x)$ with $f : \mathbf{R}^{n+1} \to \mathbf{R}^n$, $|t - t_0| \leq a$, $x \in D \subset \mathbf{R}^n$; $f(t, x)$ satisfies the Lipschitz condition with respect to $x$ if in $[t_0 - a, t_0 + a] \times D$ we have

$$\|f(t, x_1) - f(t, x_2)\| \leq L\|x_1 - x_2\|$$

with $x_1, x_2 \in D$ and $L$ a constant. $L$ is called the Lipschitz constant.

Instead of saying that $f(t, x)$ satisfies the Lipschitz condition one often uses the expression : $f(t, x)$ is Lipschitz continuous in $x$. Note that Lipschitz continuity in $x$ implies continuity in $x$.

It is easy to see that continuous differentiability implies Lipschitz continuity; this provides us in many applications with a simple criterion to establish whether a vector function satisfies this condition.

The Lipschitz condition plays an essential part in the following theorem.

**Theorem 1.1**

Consider the initial value problem

$$\dot{x} = f(t, x), \quad x(t_0) = x_0$$

with $x \in D \subset \mathbf{R}^n$, $|t - t_0| \leq a$; $D = \{x \mid \|x - x_0\| \leq d\}$, $a$ and $d$ are positive constants. The vector function $f(t, x)$ satisfies the following conditions:

a. $f(t, x)$ is continuous in $G = [t_0 - a, t_0 + a] \times D$;

b. $f(t, x)$ is Lipschitz continuous in $x$.

Then the initial value problem has one and only one solution for $|t - t_0| \leq \inf\left(a, \frac{d}{M}\right)$ with

$$M = \sup_G \|f\|.$$

**Proof**

The reader should consult an introductory text in ordinary differential equation or the book by Coddington and Levinson (1955) or Walter (1976).  □

The solution of the initial value problem will be indicated in the sequel by $x(t)$, sometimes by $x(t; x_0)$ or $x(t; t_0, x_0)$. Note that theorem 1.1 guarantees the existence of the solution in a neighbourhood of $t = t_0$, the size of which depends on the supnorm $M$ of the vector function $f(t, x)$. Often one can continue the solution outside this neighbourhood; in many problems we shall not give $a$ and $d$ apriori but instead, while analysing the problem, we shall determine the time interval of existence and the domain $D$.

## Example 1.1

$$\dot{x} = x, \quad x(0) = 1, t \geq 0.$$

The solution exists for $0 \leq t \leq a$ with $a$ an arbitrary positive constant. The solution, $x(t) = e^t$, can be continued for all positive $t$ and it becomes unbounded for $t \to \infty$.

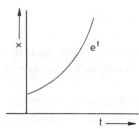

*Figure 1.1*

## Example 1.2

$$\dot{x} = x^2, \quad x(0) = 1, 0 \leq t < \infty.$$

We find $x(t) = (1-t)^{-1}$ so the solution exists for $0 \leq t < 1$. In this case we know the solution explicitly; in estimating the time interval of existence we could take $a = \infty$, $d$ some positive constant. Then $M = (1+d)^2$ and theorem 1.1 guarantees existence and uniqueness for $0 \leq t \leq \inf(\infty, d(d+1)^{-2}) = d(d+1)^{-2}$. If, for example, we are interested in values of $x$ with upperbound 3, then $d =| 3 - x_0 |= 2$. Theorem 1.1 guarantees the existence of the solution for $0 \leq t \leq 2/9$. The actual solution reaches the value 3 for $t = 2/3$.

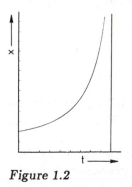

*Figure 1.2*

## 1.3 Gronwall's inequality

In parts of the theory we shall use an inequality introduced by Gronwall in 1918. The applications to the theory of differential equations are of later date and we should mention the name of Richard Bellman in this respect. Keeping an eye on our future use of the inequality we shall present here two versions of Gronwall's inequality; there are more versions and extensions.

**Theorem 1.2 (Gronwall)**
Assume that for $t_0 \leq t \leq t_0 + a$, with $a$ a positive constant, we have the estimate

(1.4)
$$\phi(t) \leq \delta_1 \int_{t_0}^{t} \psi(s)\phi(s)\,ds + \delta_3$$

in which, for $t_0 \leq t \leq t_0 + a$, $\phi(t)$ and $\psi(t)$ are continuous functions, $\phi(t) \geq 0$ and $\psi(t) \geq 0$; $\delta_1$ and $\delta_3$ are positive constants. Then we have for $t_0 \leq t \leq t_0 + a$

$$\phi(t) \leq \delta_3 e^{\delta_1 \int_{t_0}^{t} \psi(s)\,ds}$$

**Proof**
From the estimate 1.4 we derive

$$\frac{\phi(t)}{\delta_1 \int_{t_0}^{t} \psi(s)\phi(s)\,ds + \delta_3} \leq 1.$$

Multiplication with $\delta_1 \psi(t)$ and integration yields

$$\int_{t_0}^{t} \frac{\delta_1 \psi(s)\phi(s)\,ds}{\delta_1 \int_{t_0}^{s} \psi(\tau)\phi(\tau)\,d\tau + \delta_3} \leq \delta_1 \int_{t_0}^{t} \psi(s)\,ds$$

so that

$$ln\left(\delta_1 \int_{t_0}^{t} \psi(s)\phi(s)\,ds + \delta_3\right) - ln\delta_3 \leq \delta_1 \int_{t_0}^{t} \psi(s)\,ds$$

which produces

$$\delta_1 \int_{t_0}^{t} \psi(s)\phi(s)\,ds + \delta_3 \leq \delta_3 e^{\delta_1 \int_{t_0}^{t} \psi(s)\,ds}.$$

Applying the estimate 1.4 again, but now to the lefthand side, yields the required inequality. □
It is an interesting exercise to demonstrate that if $\delta_3 = 0$, the estimate 1.4 implies that $\phi(t) = 0, t_0 \leq t \leq t_0 + a$.
Sometimes a modified version is convenient.

**Theorem 1.3**
Assume that for $t_0 \leq t \leq t_0 + a$, with $a$ a positive constant, we have the estimate

(1.5)
$$\phi(t) \leq \delta_2(t - t_0) + \delta_1 \int_{t_0}^{t} \phi(s)\,ds + \delta_3$$

in which for $t_0 \leq t \leq t_0 + a$   $\phi(t)$ is a continuous function, $\phi(t) \geq 0$; $\delta_1, \delta_2, \delta_3$ are constants with $\delta_1 > 0$, $\delta_2 \geq 0$, $\delta_3 \geq 0$. Then we have for $t_0 \leq t \leq t_0 + a$

$$\phi(t) \leq \left( \frac{\delta_2}{\delta_1} + \delta_3 \right) e^{\delta_1(t-t_0)} - \frac{\delta_2}{\delta_1}.$$

**Proof**

Put $\phi(t) = \psi(t) - \frac{\delta_2}{\delta_1}$, then the estimate (1.5) becomes

(1.6) $$\psi(t) \leq \delta_1 \int_{t_0}^{t} \psi(s)\,ds + \frac{\delta_2}{\delta_1} + \delta_3.$$

Application of theorem 1.2 yields

$$\psi(t) \leq \left( \frac{\delta_2}{\delta_1} + \delta_3 \right) e^{\delta_1(t-t_0)}$$

Replacing $\psi(t)$ by $\phi(t) + \delta_2/\delta_1$ produces the required result. In the special case that $\delta_2 = \delta_3 = 0$ we have $\phi(t) = 0$ for $t_0 \leq t \leq t_0 + a$ (this case is left to the reader).   □

# 2 Autonomous equations

In this chapter we shall consider equations, in which the independent variable $t$ does not occur explicitly:

$$(2.1) \qquad\qquad \dot{x} = f(x)$$

A vector equation of the form (2.1) is called autonomous. A scalar equation of order $n$ is often written as

$$(2.2) \qquad\qquad x^{(n)} + F(x^{(n-1)}, \ldots, x) = 0$$

in which $x^{(k)} = d^k x/dt^k$, $k = 0, 1, \ldots, n$, $x^{(0)} = x$

In characterising the solutions of autonomous equations we shall use three special sets of solutions: *equilibrium* or *stationary solutions, periodic solutions* and *integral manifolds*.

## 2.1 Phase-space, orbits

We start with a simple but important property of autonomous equations.

**Lemma 2.1 (translation property)**
Suppose that we have a solution $\phi(t)$ of equation (2.1) (or (2.2)) in the domain $D \subset \mathbb{R}^n$, then $\phi(t - t_0)$ with $t_0$ a constant is also a solution.

**Proof**
Transform $t \to \tau$ with $\tau = t - t_0$. Apart from replacing $t$ by $\tau$, equation (2.1) does not change as $t$ does not occur explicitly in the righthand side. We have $\phi(t)$ is a solution of (2.1), so $\phi(\tau)$ is a solution of the transformed equation. $\qquad\square$

It follows from lemma 2.1, that if the initial value problem $\dot{x} = f(x), x(0) = x_0$ has the solution $\phi(t)$, the initial value problem $\dot{x} = f(x), x(t_0) = x_0$ has the solution $\phi(t - t_0)$. This second solution arises simply by translation along the time-axis.
For example, if $\sin t$ is a solution of equation (2.2), then also $\cos t$ is a solution. The reason is that we obtain $\cos t$ from $\sin t$ by the transformation $t \to t - \frac{1}{2}\pi$.

**Remark**
The solutions $\phi(t)$ and $\phi(t - t_0)$ which we have obtained are different. However, in phase space, which we shall discuss hereafter, these solutions correspond with the same orbital curves.

The translation property is important for the study of periodic solutions and for the theory of dynamical systems.

Consider again equation (2.1) $\dot{x} = f(x)$ with $x \in D \subset \mathbb{R}^n$. $D$ is called phase-space and for autonomous equations it makes sense to study this space separately. We start with a simple example.

**Example 2.1**

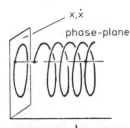

*Figure 2.1*

Consider the harmonic equation

$$\ddot{x} + x = 0.$$

The equation is autonomous; to obtain the corresponding vector equation we put $x = x_1$, $\dot{x} = x_2$ to obtain

$$\dot{x}_1 = x_2$$
$$\dot{x}_2 = -x_1$$

As we know, the solutions of the scalar equation are linear combinations of $\cos t$ and $\sin t$. It is easy to sketch the solution space $G = \mathbb{R} \times \mathbb{R}^2$. The solutions can be projected on the $x, \dot{x}$-plane which we shall call the phase-plane.

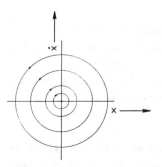

*Figure 2.2*

As time does not occur explicitly in equation (2.1), we can carry out this projection for the solutions of the general autonomous equation (2.1). The space in which we describe the behaviour of the variables $x_1, \ldots, x_n$, parametrised by $t$, is called phase-space.

A point in phase-space with coordinates $x_1(t), x_2(t), \ldots, x_n(t)$ for certain $t$, is called a phase-point. In general, for increasing $t$, a phase-point shall move through phase-space.

In carrying out the projection into phase-space, we do not generally know the solution curves of equation (2.1), However, it is simple to formulate a differential equation describing the behaviour of the orbits in phase-space.

Equation (2.1), written out in components becomes

$$\dot{x}_i = f_i(x) \quad , i = 1, \ldots, n.$$

We shall use one of the components of $x$, for instance $x_1$, as a new independent variable; this requires that $f_1(x) \neq 0$. With the chain rule we obtain $(n-1)$ equations:

(2.3)
$$\frac{dx_2}{dx_1} = \frac{f_2(x)}{f_1(x)}$$
$$\vdots \qquad \vdots$$
$$\frac{dx_n}{dx_1} = \frac{f_n(x)}{f_1(x)}$$

Solutions of system (2.3) in phase space are called orbits. If the existence and uniqueness theorem 1.1 applies to the autonomous equation (2.1) (or (2.2)), it also applies to system (2.3), describing the behaviour of the orbits in phase-space. *This means that orbits in phase space will not intersect.*

Of course we excluded the singularities of the righthand side of system (2.3) corresponding with the zeros of $f_1(x)$. If $f_2(x)$ does not vanish in these zeros, we can take $x_2$ as an independent variable, interchanging the roles of $f_1(x)$ and $f_2(x)$. If zeros of $f_1(x)$ and $f_2(x)$ coincide, we can take $x_3$ as independent variable, etc. Real problems with this construction arise in points $a = (a_1, \ldots, a_n)$ such that

$$f_1(a) = f_2(a) = \ldots = f_n(a) = 0.$$

The point $a \in \mathbf{R}^n$ is a zero of the vector function $f(x)$ and we shall call it a critical point; sometimes it is called an equilibrium point.

**Example 2.2**

Harmonic oscillator $\ddot{x} + x = 0$.

The equivalent vector equation is with $x = x_1, \dot{x} = x_2$

$$\dot{x}_1 = x_2$$
$$\dot{x}_2 = -x_1$$

Phase-space is two dimensional and the orbits are described by the equation

$$\frac{dx_2}{dx_1} = -\frac{x_1}{x_2}.$$

Integration yields

$$x_1^2 + x_2^2 = c \quad (c \text{ a constant}),$$

a family of circles in the phase-plane; $(0,0)$ is a critical point and is called a centre, see chapter 3 and figures 2.2 and 3.9.

**Example 2.3**

The equation $\ddot{x} - x = 0$.

As in the preceding example we find the equation for the orbits in the phase-plane, in this case

$$\frac{dx_2}{dx_1} = \frac{x_1}{x_2}$$

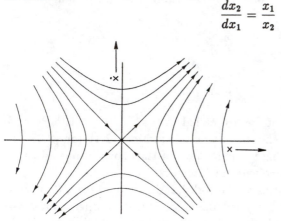

Figure 2.3

Integration produces the family of hyperboles

$$x_1^2 - x_2^2 = c \ (c \text{ a constant});$$

the critical point $(0,0)$ is called a saddle, see chapter 3.1.

The arrows indicate the direction of the motion of the phase points with time; the motion of a set of phase points along the corresponding orbits is called the *phase flow*.

## 2.2  Critical points and linearisation

Again we consider equation (2.1) $\dot{x} = f(x)$, $x \in D \subset \mathbb{R}^n$ and we assume that the vector function $f(x)$ has a zero $x = a$ in $D$.

**Definition**

The point $x = a$ with $f(a) = 0$ is called a critical point of equation $\dot{x} = f(x)$.

In the older literature, a critical point is sometimes referred to as a 'singular point'. A critical point of the equation in phase-space can be considered as an orbit, degenerated into a point.

Note also that a critical point corresponds with an equilibrium solution (or stationary solution) of the equation: $x(t) = a$ satisfies the equation for all time.

It follows from the existence and uniqueness theorem 1.1, that an equilibrium solution can never be reached in a finite time (if an equilibrium solution could be reached in a finite time, two solutions would intersect).

**Example 2.4**

$\dot{x} = -x, t \geq 0$.

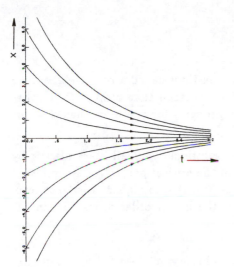

*Figure 2.4*

$x = 0$ is a critical point, $x(t) = 0$, $t \geq 0$ is equilibrium solution. Note that for all solutions starting in $x_0 \neq 0$ at $t = 0$ we have $\lim_{t \to \infty} x(t) = 0$ (the solutions are $x(t) = x_0 e^{-t}$).

**Example 2.5**

$\dot{x} = -x^2, t \geq 0$.

$x = 0$ is a critical point, $x(t) = 0$, $t \geq 0$ is equilibrium solution. The solutions starting in $x_0 \neq 0$ at $t = 0$ show qualitative and quantitative different behaviour for $x_0 > 0$ and for $x_0 < 0$. This is clear from the solutions $x(t) = (x_0^{-1} + t)^{-1}, x_0 \neq 0$. If $x_0 < 0$, the solutions become unbounded in a finite time.

In example 2.4 the solutions tend in the limit for $t \to \infty$ towards the equilibrium solution, the orbits in one-dimensional phase space tend towards the critical point. We call this phenomenon *attraction*.

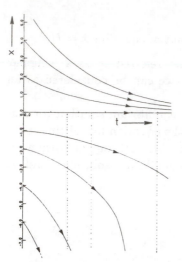

*Figure 2.5*

**Definition**

A critical point $x = a$ of the equation $\dot{x} = f(x)$ in $\mathbb{R}^n$ is called a *positive attractor* if there exists a neighbourhood $\Omega_a \subset \mathbb{R}^n$ of $x = a$ such that $x(t_0) \in \Omega_a$ implies $\lim_{t \to \infty} x(t) = a$. If a critical point $x = a$ has this property for $t \to -\infty$, then $x = a$ is called a *negative attractor*.

In analysing critical points and equilibrium solutions we shall start always by linearising the equation in a neighbourhood of the critical point. We assume that $f(x)$ has a Taylor series expansion of the first degree plus higher order rest term in $x = a$; linearising means that we leave out the higher order terms. So, in the case of

$$\dot{x} = f(x)$$

we write in the neighbourhood of the critical point $x = a$

$$\dot{x} = \frac{\partial f}{\partial x}(a)\,(x - a) + \text{higher order terms.}$$

We shall study the linear equation with constant coefficients

$$\dot{y} = \frac{\partial f}{\partial y}(a)\,(y - a).$$

To simplify the notation the point $a$ is often shifted to the origin of phase-space. Note that putting $\bar{y} = y - a$ yields

$$\dot{\bar{y}} = \frac{\partial f}{\partial y}(a)\bar{y}.$$

Often we shall abbreviate $\frac{\partial f}{\partial y}(a) = A$, a $n \times n$ matrix with constant coefficients and we omit the bar. So the linearised system which we shall study in a neighbourhood of $x = a$ is of the form

$$\dot{y} = Ay$$

**Example 2.6**

The mathematical pendulum.

The equation is $\ddot{x} + \sin x = 0$ with $-\pi \leq x \leq +\pi, x \in \mathbb{R}$ ($x$ is the angular variable indicating the deviation from the vertical).

In vector form the equation becomes with $x = x_1, \dot{x} = x_2$

$$\begin{aligned} \dot{x}_1 &= x_2 \\ \dot{x}_2 &= -\sin x_1 \end{aligned}$$

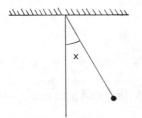

*Figure 2.6*

Critical points are $(x_1, x_2) = (0,0), (-\pi, 0), (\pi, 0)$. Expansion in a neighbourhood of $(0,0)$ yields

$$\begin{aligned} \dot{x}_1 &= x_2 \\ \dot{x}_2 &= -x_1 \quad + \text{higher order.} \end{aligned}$$

Expansion in a neighbourhood of $(\pm\pi, 0)$ yields

$$\begin{aligned} \dot{x}_1 &= x_2 \\ \dot{x}_2 &= (x_1 \mp \pi) \quad + \text{higher order.} \end{aligned}$$

Note that the phase-flow of the equation, linearised in a neighbourhood of the critical points, has been described already by examples 2.1-2.3 of section 2.1.

**Example 2.7.** The Volterra-Lotka equations

Consider the system

$$\begin{aligned} \dot{x} &= ax - bxy \\ \dot{y} &= bxy - cy \end{aligned}$$

with $x, y \geq 0$ and $a, b, c$ positive constants. This system was formulated by Volterra and Lotka to describe the interaction of two species, where $x$ denotes the population density of the prey, $y$ the population density of the predator. In this model the survival of the predators depends completely on the presence of prey; to put it mathematically, if $x(0) = 0$ than we have $\dot{y} = -cy$ so that $y(t) = y(0) \exp.(-ct)$ and $\lim\limits_{t\to\infty} y(t) = 0$.

The equilibrium solutions correspond with the critical points $(0,0)$ (no animals present, trivial) and $(c/b, a/b)$. To give an explicit example we write down $f(x,y)$ in equation 2.1 and $\partial f(x,y)/\partial(x,y)$ :

$$f(x,y) = \begin{pmatrix} ax - bxy \\ bxy - cy \end{pmatrix}, \frac{\partial f(x,y)}{\partial(x,y)} = \begin{pmatrix} a - by & -bx \\ by & bx - c \end{pmatrix}$$

Linearisation in a neighbourhood of $(0,0)$ produces, with dots for the higher order terms,

$$\begin{aligned} \dot{x} &= & ax & +\dots \\ \dot{y} &= & -cy & +\dots \end{aligned}$$

The solutions of the linearised system are of the form $x(0) \exp.(at)$ and $y(0) \exp.(-ct)$.

In a neighbourhood of $(c/b, a/b)$ we find

$$\begin{aligned} \dot{x} &= -c(y - \tfrac{a}{b}) &+\cdots \\ \dot{y} &= a(x - \tfrac{c}{b}) &+\cdots \end{aligned}$$

The solutions of the linearised system are combinations of $\cos(\sqrt{act})$ and $\sin(\sqrt{act})$.

## 2.3 Periodic solutions

**Definition**
Suppose that $x = \phi(t)$ is a solution of the equation $\dot{x} = f(x)$, $x \in D \subset \mathbb{R}^n$ and suppose there exists a positive number $T$ such that $\phi(t + T) = \phi(t)$ for all $t \in \mathbb{R}^n$. Then $\phi(t)$ is called a periodic solution of the equation with period $T$.

**Remarks**
If $\phi(t)$ has period $T$, than the solution has also period $2T$, $3T$, etc. Suppose $T$ is the smallest period, than we call $\phi(t)$ $T$-periodic. Some authors are considering the limit case of an equilibrium solution with arbitrary period $T \geq 0$, also as periodic; we shall take $T > 0$ and fixed, unless explicitly stated that the case of an equilibrium solution is also admitted.

Consider phase-space corresponding with the autonomous equation 2.1. For a periodic solution we have that after a time $T$ $\quad x = \phi(t)$ assumes the same value in $\mathbb{R}^n$. So a periodic solution produces a *closed orbit* or *cycle* in phase-space. We shall show that we can reverse this statement.

**Lemma 2.2**

A periodic solution of the autonomous equation 2.1 $\dot{x} = f(x)$ corresponds with a closed orbit (cycle) in phase-space and a closed orbit corresponds with a periodic solution.

**Proof**

We have shown already that a periodic solution produces a closed orbit in phase space. Consider now a closed orbit $C$ in phase space and a point $x_0 \in C$. The solution of equation 2.1, which we call $\phi(t)$, starts at $t = 0$ in $x_0$ and traces the orbit $C$. Because of uniqueness of solutions, $C$ cannot contain a critical point, so $\|f(x)\| \geq a > 0$ for $x \in C$. This implies that $\|\dot{x}\| \geq a > 0$ so that at a certain time $t = T$, we have returned to $x_0$. Now we want to prove that $\phi(t + T) = \phi(t)$ for all $t \in \mathbb{R}$. Note that we can write $t = nT + t_1$ with $n \in \mathbb{Z}$ and $0 < t_1 < T$. It follows from the translation property (lemma 2.1) that if $\phi(t)$ is a solution with $\phi(t_1) = x_1$, also $\phi(t - nT)$ is a solution with $\phi(t_1 + nT) = x_1$. So $\phi(t_1) = \phi(t_1 + nT)$ and as $t_1$ can be any value in $(0, T)$ we have that $\phi(t)$ is $T$-periodic. $\qquad\square$

We have seen already a linear example where closed orbits exist: the equation of the harmonic oscillator. We now give some nonlinear examples.

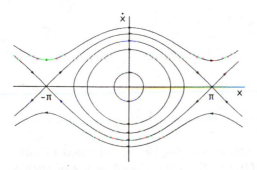

*Figure 2.7*

**Example 2.8**

The mathematical pendulum can be described by the equation

$$\ddot{x} + \sin x = 0$$

The phase-plane contains a family of closed orbits corresponding with periodic solutions.

**Example 2.9**

The van der Pol-equation

$$\ddot{x} + x = \mu(1 - x^2)\dot{x}, \ \mu > 0.$$

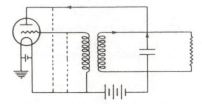

*Figure 2.8. Triode-circuit of the van der Pol-equation*

The Dutch fysicist Balthasar van der Pol formulated this equation around 1920 to describe oscillations in a triode-circuit. The phase-plane contains one closed orbit corresponding with a periodic solution (see chapter 4.4). Apart from describing these triode oscillations, the van der Pol-equation plays an important part in the theory of relaxation oscillations ($\mu \gg 1$, chapter 12) and in bifurcation theory (chapter 13).

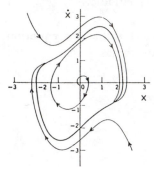

*Figure 2.9. Phase-plane of the van der Pol-equation, $\mu = 1$*

Of course the definition of a periodic solution also applies to solutions of non-autonomous equations of the form $\dot{x} = f(t, x)$. However, closed orbits of such a system do not necessarily correspond with periodic solutions as the translation property is not valid anymore. Consider for example the system

$$
\begin{aligned}
\dot{x} &= 2ty \\
\dot{y} &= -2tx
\end{aligned}
$$

with solution of the form $x(t) = \alpha \cos t^2 + \beta \sin t^2, y(t) = -\alpha \sin t^2 + \beta \cos t^2$. In the $x, y$ phase-plane we have closed orbits, the solutions are not periodic.

## 2.4   First integrals and integral manifolds

Equations 2.3, in which time has been eliminated, can be integrated in a number of cases, producing a relation between the components of the solution vector. For instance in the case of the harmonic oscillator $\ddot{x} + x = 0$, example 2.2, we have

found
$$\frac{1}{2}(\dot{x})^2 + \frac{1}{2}x^2 = E,$$

$E \geq 0$ a constant, determined by the initial conditions. We call this expression a first integral of the harmonic oscillator equation. In phase-space, this relation corresponds for $E > 0$ with a manifold, a circle around the origin. To verify that $x(t)$ and $\dot{x}(t)$ satisfy this relation we differentiate it

$$\dot{x}\ddot{x} + x\dot{x} = \dot{x}(\ddot{x} + x) = 0,$$

where we used that $x(t)$ solves the equation.

To define more generally "first integral", we introduce the concept of orbital derivative.

**Definition**
Consider the differentiable function $F : \mathsf{R}^n \to \mathsf{R}$ and the vector function $x : \mathsf{R} \to \mathsf{R}^n$. The derivative $L_t$ of the function $F$ along the vector function $x$, parameterised by $t$, is
$$L_t F = \frac{\partial F}{\partial x}\dot{x} = \frac{\partial F}{\partial x_1}\dot{x}_1 + \frac{\partial F}{\partial x_2}\dot{x}_2 + \cdots + \frac{\partial F}{\partial x_n}\dot{x}_n$$
where $x_1, \ldots, x_n$ are the components of $x$. $L_t$ is called the *orbital derivative*.

Now we choose for $x$ solutions of the differential equation 2.1 to compute the orbital derivative.

**Definition**
Consider the equation $\dot{x} = f(x), x \in D \subset \mathsf{R}^n$; the function $F(x)$ is called first integral of the equation if in $D$ holds

$$L_t F = 0$$

It follows from the definition that the first integral $F(x)$ is constant along a solution. This is why first integrals are called sometimes "constants of motion". On the other hand, taking $F(x) = $ constant we are considering the level sets of the function $F(x)$ and these level sets contain orbits of the equation. Such a level set defined by $F(x)$=constant which consists of a family of orbits is called an *integral manifold*. It will be clear that if we can find integral manifolds of an equation it helps us in understanding the build-up of phase-space of this equation.

Closely related to the concept of integral manifold is the concept of invariant set, which turns out to be very useful in the development of the theory.

**Definition**
Consider the equation $\dot{x} = f(x)$ in $D \subset \mathsf{R}^n$. The set $M \subset D$ is invariant if the

solution $x(t)$ with $x(0) \in M$ is contained in $M$ for $-\infty < t < +\infty$. If this property is valid only for $t \geq 0 (t \leq 0)$ then $M$ is called a positive (negative) invariant set.

It will be clear that equilibrium points and in general solutions which exist for all time are (trivial) examples of invariant sets. Important examples are the integral manifolds which we discussed earlier.

**Example 2.10**
The harmonic equation $\ddot{x} + x = 0$ has a first integral $F(x, \dot{x}) = \frac{1}{2}(\dot{x})^2 + \frac{1}{2}x^2$. The family of circles $F(x, \dot{x}) = E, E \geq 0$, is the corresponding set of integral manifolds.
**Example 2.11**
The second order equation $\ddot{x} + f(x) = 0$ with $f(x)$ sufficiently smooth.
After multiplication with $\dot{x}$ we can write the resulting equation as

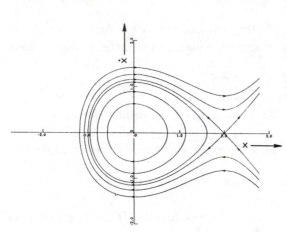

Figure 2.10 Phase plane if $f(x) = x - \frac{1}{2}x^2$

$$\frac{d}{dt}\left(\frac{1}{2}\dot{x}^2\right) + \frac{d}{dt}\int^x f(\tau)d\tau = 0.$$

So $F(x, \dot{x}) = \frac{1}{2}\dot{x}^2 + \int^x f(\tau)d\tau$ is a first integral of the equation. The orbits correspond with the level curves $F(x, \dot{x}) = $ constant. For example if $f(x) = \sin x$ (mathematical pendulum) we have

$$F(x, \dot{x}) = \frac{1}{2}\dot{x}^2 - \cos x$$

If we take $f(x) = x - \frac{1}{2}x^2$ we find

$$F(x, \dot{x}) = \frac{1}{2}\dot{x}^2 + \frac{1}{2}x^2 - \frac{1}{6}x^3.$$

The reader should analyse the critical points in the last case by linear analysis.

## Example 2.12

The equation of example 2.11 can be considered as a particular case of Hamilton's equations

$$\dot{p}_i = -\frac{\partial H}{\partial q_1}, \dot{q}_i = \frac{\partial H}{\partial p_i}, i = 1,\ldots,n$$

in which $H$ is a $C^2$ function of the $2n$ variables $p_i$, $q_i$; $H$ is a first integral of Hamilton's equations as

$$L_t H = \Sigma_{i=1}^n [\frac{\partial H}{\partial p_i}\dot{p}_i + \frac{\partial H}{\partial q_i}\dot{q}_i] = 0$$

$H$ is often called the energy integral, which applies to the cases where $H$ is the energy of a mechanical system, the dynamics of which is described by Hamilton's equations of motion. In many fields of application, the relevant equations can be written in this form.

In example 2.11 we have $n = 1$, $p_1 = \dot{x}$, $q_1 = x$ and $F(x,\dot{x})$ is written as $H(p_1,q_1)$. In the case $n = 1$, the analysis of the level sets $H(p,q) = $ constant yields a complete picture of the orbits in the phase-plane. In the case $n \geq 2$, phase space has dimension $2n \geq 4$. A study of the family of integral manifolds $H(p,q) = $ constant tells us something about the behaviour of the orbits. On such an integral manifold we still have $(2n-1)$-dimensional motion with $2n-1 \geq 3$; we shall return to these problems in chapter 15.

The analysis of level sets $H = $ constant is in general not so simple even in the two-dimensional case $(n = 1)$. Often we are interested in local behaviour of the level sets for instance in the neighbourhood of critical points. The Morse lemma is a useful tool for this.

We start with the concept of a non-degenerate critical point of a function.

## Definition

Consider $F : R^n \to R$ which is supposed to be $C^\infty$; for $x = a$ we have $\partial F/\partial x(a) = 0$ and $x = a$ is called a critical point. The point $x = a$ is called non-degenerate critical point of the function $F(x)$ if we have for the determinant

$$|\frac{\partial^2 F(a)}{\partial x^2}| \neq 0.$$

Note that $\partial F/\partial x = (\partial F/\partial x_1, \partial F/\partial x_2,\ldots,\partial F/\partial x_n)$ is a vector function, the so-called gradient of the function $F(x)$; in agreement with the notation of section 1.1, $\partial^2 F/\partial x^2$ is a matrix. The requirement of non-degeneracy we can interpret as the requirement that the vector function $\partial F/\partial x$ has a "non-degenerate linearisation" in a neighbourhood of $x = a$.

**Example 2.13** $(n = 2)$

The origin is non-degenerate critical point of the functions

$$x_1^2 + x_2^2 \quad \text{and} \quad x_1^2 - 2x_2^2.$$

The origin is degenerate critical point of the functions

$$x_1^2 x_2^2 \quad \text{and} \quad x_1^2 + x_2^3.$$

**Definition**

If $x = a$ is a non-degenerate critical point of the $C^\infty$ function $F(x)$, then $F(x)$ is called a *Morse-function* in a neighbourhood of $x = a$.

We shall discuss now the remarkable result, that the behaviour of a Morse-function in a neighbourhood of the critical point $x = a$ is determined by the quadratic part of the Taylor expansion of the function. Suppose that $x = 0$ is a non-degenerate critical point of the Morse-function $F(x)$ with expansion

$$F(x) = F_0 - c_1 x_1^2 - c_2 x_2^2 - \ldots - c_k x_k^2 + c_{k+1} x_{k+1}^2 + \ldots + c_n x_n^2 + \text{higher order terms}$$

with positive coefficients $c_1, \ldots, c_n$; $k$ is called the *index* of the critical point. There exists a transformation $x \rightarrow y$ in a neighbourhood of the critical point such that $F(x) \rightarrow G(y)$ where $G(y)$ is a Morse-function with critical point $x = 0$, the same index $k$, and apart from $G(0)$ only quadratic terms.

**Lemma 2.3 (Morse)**

Consider the $C^\infty$ function $F : \mathbb{R}^n \rightarrow \mathbb{R}$ with non-degenerate critical point $x = 0$, index k. In a neighbourhood of $x = 0$ there exists a diffeomorphism (abbreviation for transformation which is one-to-one, unique, $C^1$ and of which the inverse exists and is also $C^1$) which transforms $F(x)$ to the form

$$G(y) = G(0) - y_1^2 - y_2^2 - \ldots - y_k^2 + y_{k+1}^2 + \ldots + y_n^2.$$

**Proof**

See appendix 1. □

If we have $k = 0$, this implies that the level sets in a neighbourhood of the critical point correspond with a positive definite quadratic form; the level sets are locally diffeomorphic to a sphere. In the case of dimension $n = 2$ we have in this case closed orbits around the critical point.

**Example 2.14.**

The equation $\ddot{x} + f(x) = 0$ with $f(x)$ a $C^\infty$ function. With $x = x_1$, $\dot{x} = x_2$, the corresponding vector equation is

$$\dot{x}_1 = x_2$$
$$\dot{x}_2 = -f(x_1).$$

If $f(a) = 0$, $(a, 0)$ is a critical point; we translate $a$ to zero so we take $f(0) = 0$. Linearisation produces

$$\begin{aligned} \dot{x}_1 &= x_2 \\ \dot{x}_2 &= -f_x(0)x_1 + \text{higher order terms} \end{aligned}$$

with eigenvalues of the linear part

$$\lambda_{1,2} = \pm\sqrt{-f_x(0)}.$$

So $(0,0)$ is a centre $(f_x(0) > 0)$ or a saddle point $(f_x(0) < 0)$ for the linearised equations; we do not consider here the degenerate case $f_x(0) = 0$ (see section 13.4). Compare for the names of these critical points also the examples 2.2 and 2.3 in section 2.1; we shall return to this in a systematic way in chapter 3.
A nonlinear characterisation of the critical point can be given with the Morse-lemma. In example 2.11 we found the first integral

$$F(x_1, x_2) = \frac{1}{2}x_2^2 + \int_0^{x_1} f(\tau)d\tau.$$

$F(x_1, x_2)$ is in a neighbourhood of $(0,0)$ a Morse-function with the expansion

$$F(x_1, x_2) = \frac{1}{2}x_2^2 + \frac{1}{2}f_x(0)x_1^2 + \text{higher order terms.}$$

If $f_x(0) > 0$, there exists a $C^1$ transformation to the quadratic form $y_2^2 + y_1^2$ (index $k = 0$) so we have also in the nonlinear system closed orbits (nonlinear centre); if $f_x(0) < 0$, the transformation according to Morse produces $y_2^2 - y_1^2$ (index 1, saddle).
The reader should apply these results to the equation for the mathematical pendulum $\ddot{x} + \sin x = 0$.

## Example 2.15
The Volterra-Lotka equations.
Consider again the system in example 2.7

$$\begin{aligned} \dot{x} &= ax - bxy \\ \dot{y} &= bxy - cy \end{aligned}$$

with $x, y \geq 0$ and positive parameters $a, b$ and $c$. The equation for the orbits in the phase-plane is

$$\frac{dx}{dy} = \frac{x}{y}\frac{a - by}{bx - c}.$$

We can integrate this equation by separation of variables; we find for $x, y > 0$

$$bx - c\ln x + by - a\ln y = C$$

where the constant $C$ is determined by the initial conditions. The expression

$$F(x,y) = bx - clnx + by - alny$$

is a first integral of the Volterra-Lotka equations. In example 2.7 linearisation produced periodic solutions in a neighbourhood of the critical point $(c/b, a/b)$, corresponding with closed orbits around the critical point. Using the first integral $F(x,y)$ we can see whether this result is also valid for the complete, nonlinear equations. $F(x,y)$ is a Morse-function in a neighbourhood of $(c/b, a/b)$ with expansion

$$F(x,y) = F(c/b, a/b) + \frac{b^2}{2c}(x - \frac{c}{b})^2 + \frac{b^2}{2a}(y - \frac{a}{b})^2 + \dots$$

The critical point has index zero and using the Morse-lemma it follows that the orbits are closed.

It is interesting to note that the Volterra-Lotka equations can be put into Hamiltonian formulation as follows.

Putting $p = lnx$ , $q = lny$ the Volterra-Lotka equations becomes

$$\frac{dp}{dt} = a - be^q$$
$$\frac{dq}{dt} = be^p - c$$

Transforming the first integral we have

$$H(p,q) = be^p - cp + be^q - aq$$

so that $\dot{p} = -\partial H/\partial q$ , $\dot{q} = \partial H/\partial p$.

## 2.5 Evolution of a volume element, Liouville's theorem

We conclude this chapter by characterising the change of the volume of an element in phase space, where this change is caused by the flow of an autonomous differential equation.

**Lemma 2.4**
Consider the equation $\dot{x} = f(x)$ in $R^n$ and a domain $D(0)$ in $R^n$ which is supposed to have a volume $v(0)$. The flow defines a mapping $g$ of $D(0)$ into $R^n$, $g : R^n \to R^n$, $D(t) = g^t D(0)$. For the volume $v(t)$ of the domain $D(t)$ we have

$$\frac{dv}{dt}|_{t=0} = \int_{D(0)} \nabla.f dx$$

$(\nabla.f = \partial f_1/\partial x_1 + \partial f_2/\partial x_2 + \dots + \partial f_n/\partial x_n)$.

**Proof**
It follows from the definition of the Jacobian of a transformation that

$$v(t) = \int_{D(0)} |\frac{\partial g^t(x)}{\partial x}| dx.$$

We expand $g^t(x) = x + f(x)t + O(t^2)$ for $t \to 0$. It follows that

$$\frac{\partial g^t(x)}{\partial x} = I + \frac{\partial f}{\partial x}t + O(t^2)$$

with $I$ the $n \times n$-identity matrix and

$$|\frac{\partial g^t(x)}{\partial x}| = |I + \frac{\partial f}{\partial x}t| + O(t^2)$$

$$= 1 + Tr(\frac{\partial f}{\partial x})t + O(t^2).$$

The last step follows from the expansion of the determinant. As $Tr(\frac{\partial f}{\partial x}) = \nabla.f$ we have

$$v(t) = v(0) + \int_{D(0)} t\nabla.f \ dx + O(t^2)$$

from which follows the lemma. □

Lemma 2.4 is useful to obtain global insight in the phase flow corresponding with an autonomous equation.
In the case of example 2.4 $\dot{x} = -x$, where we have attraction towards $x = 0$, lemma 2.4 yields

$$\frac{dv}{dt}|_{t=0} = -\int_{D(0)} dx = -v(0)$$

The volume element shrinks exponentially.
In example 2.6, the mathematical pendulum, we find

$$\frac{dv}{dt}|_{t=0} = 0$$

and as $t = 0$ is taken arbitrarily, phase volume is constant under the flow. The case of the Volterra-Lotka equations, examples 2.7 and 2.15 yields

$$\frac{dv}{dt}|_{t=0} = -\int_{D(0)} (a - c - by + bx)dx \ dy.$$

From the analysis of the equations in this case, we know that a volume element periodically shrinks and expands.
An important application of lemma 2.4 is to Hamilton's equations which have been introduced in example 2.12.

## Theorem 2.1 (Liouville)

The flow generated by a time-independent hamiltonian system is volume preserving.

**Proof**

We find $\nabla.f = 0$, so that $\dot{v}(t_0) = 0$ for $t_0 = 0$ and for all $t_0 \in \mathbb{R}$. $\qquad\square$

## 2.6  Exercises

2-1. Consider the equation

$$\ddot{x} - \lambda\dot{x} - (\lambda - 1)(\lambda - 2)x = 0$$

with $\lambda$ a real parameter.

Find the critical points and characterise these points. Sketch the flow in the phase-plane and indicate the direction of the flow.

2-2. An extension of the Volterra-Lotka model of examples 2.7 and 2.15 is obtained by taking into account the saturation affect which is caused by a large number of prey. The equations become

$$\dot{x} = ax - b\frac{xy}{1 + sx}$$

$$\dot{y} = b\frac{xy}{1 + sx} - cy$$

with $x, y \geq 0$; the parameters are all positive.

Determine the equilibrium solutions and their attraction properties by linearisation.

2-3. We are studying the three-dimensional system

$$\dot{x}_1 = x_1 - x_1 x_2 - x_2^3 + x_3(x_1^2 + x_2^2 - 1 - x_1 + x_1 x_2 + x_2^3)$$
$$\dot{x}_2 = x_1 - x_3(x_1 - x_2 + 2x_1 x_2)$$
$$\dot{x}_3 = (x_3 - 1)(x_3 + 2x_3 x_2^2 + x_3^3)$$

a. Determine the critical points of this system.

b. Show that the planes $x_3 = 0$ and $x_3 = 1$ are invariant sets.

c. Consider the invariant set $x_3 = 1$. Does this set contain periodic solutions?

2-4. Suppose that a very long conductor has been fixed in a vertical straight line; a constant current $I$ passes through the conductor. A small conductor, length $l$ and mass $m$, has also been placed in a vertical straight line; it has

been fixed to a spring which can move horizontally. A constant current $i$
passes through the small conductor. See the figure:

The small conductor will be put into motion in the $x$-direction but it remains
parallel to the long conductor; without deformation of the spring, its position
is $x = 0$. The fixed position of the long conductor is given by $x = a$. The
equation of motion of the small conductor is

$$m\ddot{x} + kx - \frac{2Iil}{a - x} = 0$$

with $k$ positive and $x < a$.

a. Show that, putting $\lambda = 2Iil/k$, the equation can be written as

$$\ddot{x} - \frac{k}{m} \frac{x^2 - ax + \lambda}{a - x} = 0$$

b. Put the equation in the frame-work of Hamiltonian systems.

c. Compute the critical points and characterise them. Does the result
agree with the Hamiltonian nature of the problem?

d. Sketch the phase-plane for various values of $\lambda$.

2-5. Find the critical points of the system

$$\dot{x} = y$$
$$\dot{y} = x - 2x^3.$$

Characterise the critical points by linear analysis and determine their attraction properties.

2-6. Consider the system

$$\dot{x} = x$$
$$\dot{y} = y.$$

a. Find a first integral.

b. Can we derive the equations from a Hamilton function?

2-7. Consider in $\mathbf{R}^n$ the system $\dot{x} = f(x)$ with divergence $\nabla.f(x) = 0$. So the phase-flow is volume-preserving. Does this mean that a first integral exists; solve this question for $n = 2$.

2-8. Determine the critical points of the system

$$\dot{x} = x^2 - y^3$$
$$\dot{y} = 2x(x^2 - y).$$

Are there attractors in the system? Determine a first integral. Do periodic solutions exist?

# 3   Critical points

In section 2.2 we saw that linearisation in a neighbourhood of a critical point of an autonomous system $\dot{x} = f(x)$ leads to the equation

(3.1)
$$\dot{y} = Ay$$

with $A$ a constant $n \times n$-matrix; in this formulation the critical point has been translated to the origin. We exclude in this chapter the case of a singular matrix $A$, so

$$\det A \neq 0.$$

Put in a different way, we assume that the critical point is non-degenerate.

In analysing the critical point of the linear system we first determine the eigenvalues of $A$ from the characteristic equation

$$\det(A - \lambda I) = 0.$$

The eigenvalues are $\lambda_1, \ldots, \lambda_n$.

There exists a real, non-singular matrix $T$ such that $T^{-1}AT$ is in so-called Jordan form; for a detailed description in the context of differential equations see the books of Coddington and Levinson (1955), Hale (1969) or Walter (1976). We summarise the results.

If the $n$ eigenvalues are different, then $T^{-1}AT$ is in diagonal form with the eigenvalues as diagonal elements. If there are some equal eigenvalues, the linear transformation $y = Tz$ still leads to a serious simplification. We find

$$T\dot{z} = ATz \quad \text{or}$$

(3.2)
$$\dot{z} = T^{-1}ATz.$$

As the Jordan normal form $T^{-1}Az$ is so simple, we can integrate equation (3.2) immediately, from which $y = Tz$ follows. In characterising the critical point and the corresponding phase flow, we shall assume that we have already obtained equation 3.2 by a linear transformation.

## 3.1   Two-dimensional linear systems

We shall give the location of the eigenvalues by a diagram which consists of the complex plane (real axis horizontally, imaginary axis vertically), where the eigenvalues have been indicated by dots. In this section the dimension is two so the eigenvalues $\lambda_1$ and $\lambda_2$ are both real or complex conjugate. If $\lambda_1 \neq \lambda_2$ (real or complex) then $T^{-1}AT$ is of the form

$$\begin{pmatrix} \lambda_1 & 0 \\ 0 & \lambda_2 \end{pmatrix}$$

We find for $z(t)$ the general solution

(3.3)
$$z(t) = \begin{pmatrix} c_1 e^{\lambda_1 t} \\ c_2 e^{\lambda_2 t} \end{pmatrix}$$

with $c_1$ and $c_2$ arbitrary constants. The behaviour of the solutions represented by 3.3 is for all kind of choices of $\lambda_1$ and $\lambda_2$ very different. We have the following cases.

*Figure 3.1 Eigenvalue diagram*

a. The node.
The eigenvalues are real and have the same sign. If $\lambda_1 \neq \lambda_2$ we have with $z = (z_1, z_2)$ the real solutions $z_1(t) = c_1 e^{\lambda_1 t}$ and $z_2(t) = c_2 e^{\lambda_2 t}$. Elimination of $t$ produces $|z_1| = c|z_2|^{\lambda_1/\lambda_2}$ with $c$ a constant. So in the phase-plane we find orbits which are related to parabolas, see figure 3.2.

We call this critical point a node. If $\lambda_1, \lambda_2 < 0$ then $(0,0)$ is a positive attractor, if $\lambda_1, \lambda_2 > 0$ then $(0,0)$ is a negative attractor.
If $\lambda_1 = \lambda_2 = \lambda$, the normal form is in general

$$\begin{pmatrix} \lambda & 1 \\ 0 & \lambda \end{pmatrix}$$

so that

$$\begin{aligned} z_1(t) &= c_1 e^{\lambda t} + c_2 t e^{\lambda t} \\ z_2(t) &= c_2 e^{\lambda t} \end{aligned}$$

with $c_1$ and $c_2$ constants; see figure 3.3. In this case the normal form can also be a diagonal matrix

$$\begin{pmatrix} \lambda & 0 \\ 0 & \lambda \end{pmatrix}.$$

It is easy to verify that in the phase-plane the orbits are straight lines through the origin.

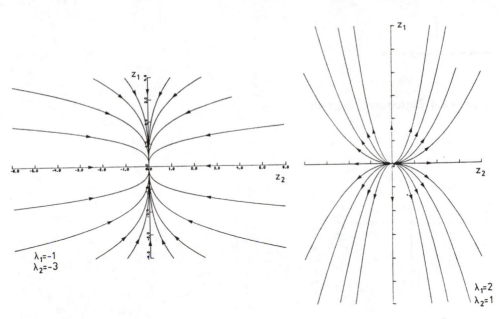

Figure 3.2

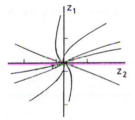

Figure 3.3

b. The saddle point.
The eigenvalues $\lambda_1$ and $\lambda_2$ are real and have different sign. The solutions are again of the form given by equation 3.3. In the phase-plane the orbits are given by

$$|z_1| = c|z_2|^{-|\lambda_1/\lambda_2|}$$

with $c$ a constant. The behaviour of the orbits is hyperbolic, the critical point $(0,0)$ is not an attractor; we shall call this a saddle point. It should be noted that the coordinate axes correspond with five different solutions: the critical point $(0,0)$ and the four half axes.

Furthermore we note that there exist two solutions with the property $(z_1(t), z_2(t)) \rightarrow (0,0)$ for $t \rightarrow \infty$ and two solutions with this property for $t \rightarrow -\infty$. The first two

*Figure 3.4 Eigenvalue diagram*

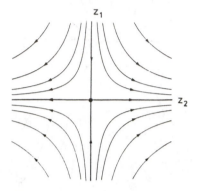

*Figure 3.5 Saddle point*

are called the *stable manifolds* of the saddle point, the other two the *unstable manifolds*.

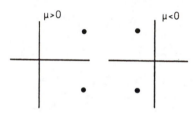

*Figure 3.6 Eigenvalue diagram*

c. *The focus*

The eigenvalues $\lambda_1$ and $\lambda_2$ are complex conjugate, $\lambda_{1,2} = \mu \pm wi$ with $\mu w \neq 0$. The complex solutions are of the form $exp.((\mu \pm wi)t)$. Linear combination of the complex solutions produces real independent solutions of the form $e^{\mu t} cos(wt)$, $e^{\mu t} sin(wt)$.

The orbits are spiralling in or out with respect to $(0,0)$ and we call $(0,0)$ a focus. In the case of spiralling in as in figure 3.7, $\mu < 0$, the critical point is a positive attractor; if $\mu > 0$ then $(0,0)$ is a negative attractor.

d. *The centre*

The special case that the eigenvalues are purely imaginary occurs quite often in applications. If $\lambda_{1,2} = \pm wi$, $w$ real, then $(0,0)$ is called a centre. The solutions can

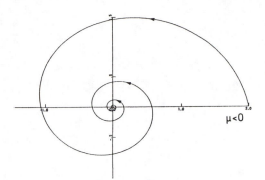

$\mu < 0$

*Figure 3.7 Focus*

*Figure 3.8 Eigenvalue diagram*

be written as a combination of $\cos(wt)$ and $\sin(wt)$, the orbits in the phase-plane are circles. It is clear that $(0,0)$ is not an attractor.

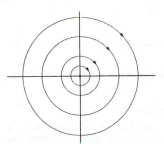

*Figure 3.9 Centre*

## 3.2 Remarks on three-dimensional linear systems

Increasing the dimension $n$ of equation 3.1 adds quickly to the number of possible cases. To illustrate such discussions we shall make some introductory remarks about the case $n = 3$. The eigenvalues $\lambda_1$, $\lambda_2$ and $\lambda_3$ can be real or one real and two complex conjugate.

If the eigenvalues are all different, as in the case of one real and two complex conjugate eigenvalues, then $T^{-1}AT$ is a diagonal matrix and we have

$$z(t) = (c_1 \ exp.(\lambda_1 t), c_2 \ exp.(\lambda_2 t), c_3 \ exp.(\lambda_3 t))$$

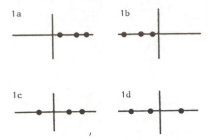

Figure 3.10 Eigenvalue diagrams

with $c_1$, $c_2$ and $c_3$ arbitrary constants. We consider the following cases:
a. Three real eigenvalues. If the eigenvalues have all the same sign $(1a - b)$ then
we call this a three-dimensional node.
The cases $1c - d$ are of type saddle-node.

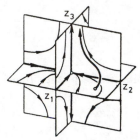

Figure 3.11 Phase space in the case 1d

Figure 3.11 shows phase space in the case 1d, where the positive eigenvalue corresponds with the vertical direction.
Identical eigenvalues produce various Jordan normal forms.
Two identical eigenvalues $\lambda_2 = \lambda_3 = \lambda$. There are two possibilities for the matrix
$T^{-1}AT$ in equation 3.2

$$\begin{pmatrix} \lambda_1 & 0 & 0 \\ 0 & \lambda & 0 \\ 0 & 0 & \lambda \end{pmatrix} \text{ and } \begin{pmatrix} \lambda_1 & 0 & 0 \\ 0 & \lambda & 1 \\ 0 & 0 & \lambda \end{pmatrix}$$

Three identical eigenvalues $\lambda_1 = \lambda_2 = \lambda_3 = \lambda$ corresponds with the cases

$$\begin{pmatrix} \lambda & 0 & 0 \\ 0 & \lambda & 0 \\ 0 & 0 & \lambda \end{pmatrix}, \begin{pmatrix} \lambda & 0 & 0 \\ 0 & \lambda & 1 \\ 0 & 0 & \lambda \end{pmatrix} \text{ and } \begin{pmatrix} \lambda & 1 & 0 \\ 0 & \lambda & 1 \\ 0 & 0 & \lambda \end{pmatrix}$$

The analysis of the phase flow in these cases is left to the reader.

Figure 3.12 Eigenvalue diagram

b. Two complex eigenvalues $\lambda_2$, $\lambda_3$ with $\lambda_1$ and $Re\ \lambda_2, \lambda_3$ the same sign. In the case 2a the critical point $(0,0)$ is a negative attractor, in the case 2b a positive attractor. In figure 3.13 we give two orbits in the case 2b; it is instructive to analyse also the case where $Re\ \lambda_2, \lambda_3 < \lambda_1 < 0$ and case 2a.

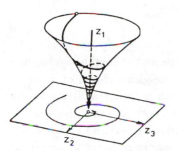

Figure 3.13  Phase space in the case 2b

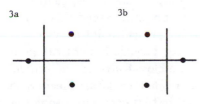

Figure 3.14 Eigenvalue diagram

c. Two complex eigenvalues $\lambda_2$ and $\lambda_3$ with $\lambda_1$ and $Re\ \lambda_2, \lambda_3$ a different sign. In the case 3a we have attraction in one direction and negative attraction (expansion) in the other two directions; the case 3b shows complementary behaviour.

d. Two eigenvalues purely imaginary. There is only in one direction positive (4b) or negative (4a) attraction.

## 3.3   Critical points of nonlinear equations

Up till now we have analysed critical points of autonomous equations $\dot{x} = f(x)$ by linear analysis. We assume that the critical point has been translated to $x = 0$

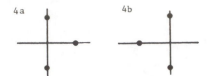

*Figure 3.15 Eigenvalue diagram*

and that we can write the equation in the form

(3.4) $$\dot{x} = Ax + g(x)$$

with $A$ a non-singular $n \times n$-matrix and that

$$\lim_{\|x\| \to 0} \frac{\|g(x)\|}{\|x\|} = 0.$$

The last assumption holds under rather general conditions, for instance when the vector function $f(x)$ is continuously differentiable in a neighbourhood of $x = 0$.
The analysis of nonlinear equations usually starts with a linear analysis as described here, after which one tries to draw conclusions about the original, nonlinear system. It turns out that some properties of the linearised system also hold for the nonlinear system, other properties do not carry over from linear to nonlinear systems. Here we shall give some examples of the relation linear-nonlinear; most of these discussions will be repeated and extended in the subsequent chapters.
At first we have the problem of the nomenclature node, saddle, focus, centre for the nonlinear equation. For plane systems $(n = 2)$ we can still extend this terminology to the nonlinear case. For instance: a node is a critical point at which all orbits arrive with a certain tangent for $t \to +\infty$ or $-\infty$; a centre is a critical point for which a neighbourhood exists where all orbits are closed around the critical point. In considering systems with dimension $n > 2$ this nomenclature meets with many problems and so we characterise a critical point in that case with properties as attraction, eigenvalues, existence of (un)stable manifolds etc.
Preambling to chapter 7 we formulate two theorems.

**Theorem 3.1**
Consider equation 3.4; if $x = 0$ is a positive (negative) attractor for the linearised equation, than $x = 0$ is a positive (negative) attractor for the non-linear equation 3.4.

At the other hand, if the critical point is a saddle, than the critical point cannot be an attractor for the nonlinear equation. More in general:

**Theorem 3.2**
Consider equation 3.4; if the matrix $A$ has an eigenvalue with positive real part, then the critical point $x = 0$ is not a positive attractor for the nonlinear equation 3.4.
The two theorems will be proved in chapter 7.

Another important result concerns the existence of stable and unstable manifolds as we found them for the saddle point. It turns out that they also exist for the nonlinear system; we generalise directly to $\mathbb{R}^n$

**Theorem 3.3 (existence of stable and unstable manifolds)**
Consider the equation

$$\dot{x} = Ax + g(x) \quad , x \in \mathbb{R}^n$$

The constant $n \times n$-matrix $A$ has $n$ eigenvalues with nonzero real part, $g(x)$ is smooth and

$$\lim_{\|x\| \to 0} \frac{\|g(x)\|}{\|x\|} = 0.$$

Then, in a neighbourhood of the critical point $x = 0$, there exist stable and unstable manifolds $W_s$ and $W_u$ with the same dimensions $n_s$ and $n_u$ as the stable and unstable manifolds $E_s$ and $E_u$ of the system

$$\dot{y} = Ay$$

In $x = 0$ $E_s$ and $E_u$ are tangent to $W_s$ and $W_u$.

**Proof**
This can be found in Hartman (1964) chapter 9 or in Knobloch and Kappel (1974) chapter 5.  □

In the literature one finds many extensions and related theorems; see also chapter 13.4 for the case where one has eigenvalues with real part zero.
An interesting result is that, under the conditions of theorem 3.3, the phase-flows of the linear and the nonlinear equations are homeomorph in a neighbourhood of $x = 0$ (a homeomorphism is an abbreviation for a continuous mapping which has a continuous inverse). It turns out that to have a diffeomorphism between the flow of the linear and the nonlinear equation, one has to put heavier requirements on equation 3.4.

**Example 3.1** Consider the system in $\mathbb{R}^2$

$$\dot{x} = -x$$
$$\dot{y} = 1 - x^2 - y^2.$$

with critical points $(0, 1)$ and $(0, -1)$. We have

$$\frac{\partial}{\partial(x, y)} \begin{pmatrix} -x \\ 1 - x^2 - y^2 \end{pmatrix} = \begin{pmatrix} -1 & 0 \\ -2x & -2y \end{pmatrix}$$

In $(0,1)$ the eigenvalues are $-1$ and $-2$, so by linear analysis we have a node which is a positive attractor. In $(0,-1)$ the eigenvalues are $-1$ and $+2$, so by linear analysis we have a saddle. In figure 3.16 the phase-flow is presented. It is easy to see that the conditions of theorem 3.3 for the existence of stable and unstable manifolds have been satisfied.

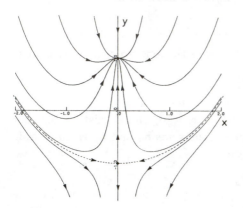

*Figure 3.16*

**Remarks**
The stable manifolds of the saddle separate the phase-plane into two domains, where the behaviour of the orbits is qualitatively different. Such a manifold we call a *separatrix*.
In numerical calculations of stable and unstable manifolds it is convenient to start in a neighbourhood of the saddle in points of $E_s$ and $E_u$, which have been obtained from the linear analysis (in the special case of example 3.1 we have $E_u = W_u$). In the case of a stable manifold we are integrating of course for $t \leq t_0$.

**Example 3.2**
If linear analysis leads to finding a centre, we need in general additional information to establish the character of the critical point for the full, non-linear equation. Examples of a type of additional information, existence of integral manifolds, have been given in section 2.4. Another type of behaviour is illustrated as follows. Consider in $\mathbf{R}^2$ the system

$$\begin{aligned} \dot{x} &= -y - ax(x^2 + y^2)^{1/2} \text{ , a constant} \\ \dot{y} &= x - ay(x^2 + y^2)^{1/2}. \end{aligned}$$

The critical point $(0,0)$ is a centre by linear analysis. We transform to polar coordinates $x, y \to r, \theta$ by $x = r\cos\theta$ , $y = r\sin\theta$ to find the system $\dot{r} = -ar^2$ , $\dot{\theta} = 1$. Solutions: $r(t) = (r(0)^{-1} + at)^{-1}$, $\theta(t) = \theta(0) + t$. If $a \neq 0$, the critical point $(0,0)$

is not a centre of the nonlinear system. If for instance $a > 0$, $(0,0)$ is a positive attractor.

**Example 3.3** (degenerate critical point)
Consider in $\mathbb{R}^2$ the system

$$\dot{x} = -x + y^2$$
$$\dot{y} = -y^2 + x^2.$$

The critical points are $(0,0)$ and $(1,1)$ of which $(1,1)$ is a saddle; $(0,0)$ has one eigenvalue zero and in the beginning of this chapter we called such a critical point degenerate. The phase flow has been depicted in figure 3.17.

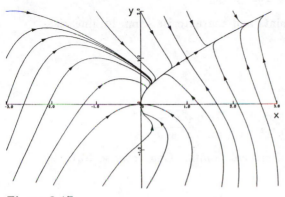

Figure 3.17

It is also interesting to compare the flow in the neighbourhood of the saddle with the saddle of figure 3.5. It appears that in the case of figure 3.17 there is a lot of difference in flow velocities in the two directions. One can illustrate this by calculating the eigenvalues.

## 3.4   Exercises

3-1. Consider the two-dimensional system in $\mathbb{R}^2$

$$\dot{x} = y(1 + x - y^2)$$
$$\dot{y} = x(1 + y - x^2)$$

Determine the critical points and characterise the linearised flow in a neighbourhood of these points.

3-2. We are studying the geometrical properties of the critical point $(0,0)$ of the autonomous system

$$\dot{x} = ax + by + f(x,y)$$
$$\dot{y} = cx + dy + g(x,y)$$

with $f(x,y) = 0(r), g(x,y) = 0(r)$ as $r = \sqrt{x^2 + y^2} \to 0$.

a. Suppose that $(0,0)$ is a focus in the linearised system. Show that in the full system, the orbits are also spiralling near $(0,0)$ i.e. for each orbit near $(0,0)$ the polar angle $\theta(t)$ takes all values in $[0, 2\pi)$ for $t > a$ and each $a \in \mathbb{R}$.

b. Can we prove a similar statement for node-like behaviour. Analyse the case

$$a = -1, b = 0, c = 0, d = -1, \ f(x,y) = -y/\log(x^2 + y^2)^{1/2},$$
$$g(x,y) = -y/\log(x^2 + y^2)^{1/2}.$$

3-3. Consider the system

$$\dot{x} = 16x^2 + 9y^2 - 25$$
$$\dot{y} = 16x^2 - 16y^2$$

a. Determine the critical points and characterise them by linearisation.

b. Sketch the phase-flow.

3-4. Consider the system

$$\dot{x} = -y + x(x^2 + y^2) \sin \sqrt{x^2 + y^2}$$
$$\dot{y} = x + y(x^2 + y^2) \sin \sqrt{x^2 + y^2}$$

a. Determine the number of critical points. Characterise $(0,0)$ by linear analysis.

b. Is $(0,0)$ an attractor?

3-5. In certain approximations one studies the equation

$$\ddot{x} + c\dot{x} - x(1 - x) = 0$$

with a special interest in solutions with the properties:

$$\lim_{t \to -\infty} x(t) = 0, \ \lim_{t \to +\infty} x(t) = 1, \ \dot{x}(t) > 0 \text{ for } -\infty < t < +\infty.$$

Derive a necessary condition for the parameter $c$ for such solutions to exist.

3-6. Establish the attraction properties of the solutions $(0,0)$ of

$$\dot{x} = x^3 + y$$
$$\dot{y} = (x^2 + y^2 - 2)y$$

3-7. Determine the critical points of the system

$$\dot{x} = x(1 - x^2 - 6y^2)$$
$$\dot{y} = y(1 - 3x^2 - 3y^2)$$

and characterise them by linear analysis.

# 4  Periodic solutions

The concept of a periodic solution of a differential equation was introduced in section 2.3. We have shown that in the case of an autonomous equation the periodic solutions correspond with closed orbits in phase-space.

Autonomous two-dimensional systems with phase-space $R^2$ are called plane (or planar) systems. Closed orbits in $R^2$, corresponding with periodic solutions, do not intersect because of uniqueness. According to the Jordan separation theorem they split $R^2$ into two parts, the interior and the exterior of the closed orbit. This topological property leads to a number of special results named after Poincaré and Bendixson. In preparing for the main theorem, section 4.3, we have collected a number of concepts and results in section 4.2 which are important in the more general setting of $R^n$. These ideas turn out to be very useful in later chapters, in particular in the theory of dynamical systems in the chapters 14 and 15.

Periodic solutions of autonomous systems with dimension larger than two or of non-autonomous systems are more difficult to analyse. In section 4.5 we shall formulate some results about systems in $R^n$; also we shall return to the subject of periodic solutions in most of the subsequent chapters.

## 4.1  Bendixson's criterion

Consider the plane autonomous system

(4.1)
$$\dot{x} = f(x,y), \ \dot{y} = g(x,y)$$

in a domain $D \subset R^2$.

**Theorem 4.1** (criterion of Bendixson)
Suppose that the domain $D \subset R^2$ is simply connected (there are no 'holes' or 'separate parts' in the domain); $(f,g)$ is continuously differentiable in $D$. Equation 4.1 can only have periodic solutions if $\nabla.(f,g)$ changes sign in $D$ or if $\nabla.(f,g) = 0$ in $D$.

**Proof**
Suppose that we have a closed orbit $C$ in $D$, corresponding with a solution of equation 4.1; the interior of $C$ is $G$. We shall apply the Gauss theorem in the form

$$\int_G \int \nabla \cdot (f,g)\, d\sigma = \int_C (f\, dy - g\, dx) = \int_C (f\frac{dy}{ds} - g\frac{dx}{ds})\, ds.$$

The integrand in the last integral is zero as the closed orbit $C$ corresponds with a solution of equation 4.1. So the integral vanishes, but this means that the divergence of $(f,g)$ cannot be sign definite. $\qquad \Box$

## Example 4.1

We know that the damped linear oscillator contains no periodic solutions. We shall consider now a nonlinear oscillator with nonlinear damping represented by the equation

(4.2) $$\ddot{x} + p(x)\dot{x} + q(x) = 0$$

We assume that $p(x)$ and $q(x)$ are smooth and that $p(x) > 0$, $x \in \mathbf{R}$ (damping). With $x = x_1$, $\dot{x} = x_2$ we have the equivalent vector equation

$$\begin{aligned} \dot{x}_1 &= x_2 \\ \dot{x}_2 &= -q(x_1) - p(x_1)x_2. \end{aligned}$$

The divergence of the vector function is $-p(x_1)$ which is negative definite. It follows from Bendixson's criterion that equation 4.2 has no periodic solutions.

## Example 4.2

Consider in $\mathbf{R}^2$ the van der Pol equation

$$\ddot{x} + x = \mu(1 - x^2)\dot{x}, \quad \mu \text{ a constant}$$

With $x = x_1$, $\dot{x} = x_2$ we have the vector equation

$$\begin{aligned} \dot{x}_1 &= x_2 \\ \dot{x}_2 &= -x_1 + \mu(1 - x_1^2)x_2. \end{aligned}$$

The divergence of the vector function is $\mu(1 - x_1^2)$. A periodic solution, if it exists, has no intersect with $x_1 = +1$, $x_1 = -1$ or both (see figure 2.9 in section 2.3).

## Example 4.3

Consider in $\mathbf{R}^2$ the system of example 3.3

$$\begin{aligned} \dot{x} &= -x + y^2 \\ \dot{y} &= -y^3 + x^2 \end{aligned}$$

We have seen that there are two critical points, $(0,0)$ which is degenerate and $(1,1)$ which is a saddle. The divergence of the vector function is $-1 - 3y^2$; as this expression is negative definite, the system has no periodic solutions.

## Remark

The theorem has been formulated and proved for vector functions in $\mathbf{R}^2$. One might wonder whether the theorem can be generalised to systems with dimension larger that 2. Unfortunately this is not the case; it is easy to find a counterexample for $n = 3$ and the reader is invited to construct such an example.

The requirements for the domain $D$ to be simple and connected are essential. If necessary, the reader should consult here an introductory text to higher dimensional integration.

## 4.2 Geometric auxiliaries, preparation for the Poincaré- Bendixson theorem

In this section we shall introduce a number of concepts which are necessary in formulating and proving the Poincaré-Bendixson theorem. This theorem holds for plane, autonomous systems. The concepts which we are introducing here, however, are meaningful in the more general setting of $R^n$ and they will play a part again in subsequent chapters.

Consider the autonomous equation

$$(4.3) \qquad\qquad \dot{x} = f(x)$$

with $x \in D \subset R^n$, $t \in R$.

In section 2.4 we introduced the concept of an invariant set $M$: a set in $R^n$ such that if $x(0)$ is contained in $M$, $x(t)$ is contained in $M$ for $t \in R$. We reformulate some of the contents of section 2.1.

A solution $x(t)$ of equation 4.3 with initial value $x(0) = x_0$, corresponds in phase space with an orbit which we indicate by $\gamma(x_0)$. So if $x(t_1) = x_1$, than we have $\gamma(x_1) = \gamma(x_0)$. Sometimes we shall distinguish between the behaviour of the solution for $t \geq 0$, corresponding with the positive orbit $\gamma^+(x_0)$, and behaviour for $t \leq 0$, corresponding with the negative orbit $\gamma^-(x_0)$; $\gamma(x_0) = \gamma^-(x_0) \cup \gamma^+(x_0)$. In the case of periodic solutions we have $\gamma^+(x_0) = \gamma^-(x_0)$.

### Definition
A point $p$ in $R^n$ is called positive limitpoint of the orbit $\gamma(x_0)$ corresponding with the solution $x(t)$ if an increasing sequence of numbers $t_1, t_2, \cdots \to \infty$ exists such that the points of $\gamma(x_0)$ corresponding with $x(t_1), x(t_2), \cdots$ have limitpoint $p$. In the same way we define negative limitpoint using a decreasing sequence of numbers.

### Example 4.4

Consider in $R^2$ the system $\dot{x}_1 = -x_1$, $\dot{x}_2 = -2x_2$. The origin is a node for the phase flow and a positive attractor. It is clear that the origin is a positive attractor and positive limitpoint for all orbits, see figure 4.1.

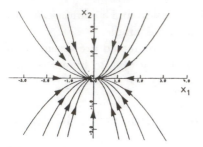

*Figure 4.1*

**Example 4.5**

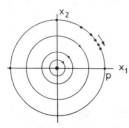

*Figure 4.2*

Consider in $\mathbf{R}^2$ the system $\dot{x}_1 = x_2$, $\dot{x}_2 = -x_1$ (harmonic oscillator). The origin is a centre, all orbits are closed. Consider an orbit $\gamma$. It is clear that each point $p$ of $\gamma$ is both positive and negative limitpoint of $\gamma$. Choose for instance $p = (1,0)$. The orbit $\gamma$ has the parametrisation $(\cos t, -\sin t)$; choose for a positive limitpoint the sequence $t_n = 2\pi n - \pi/n$, $n = 1, 2, \cdots$ so $(x_1(t_n), x_2(t_n)) \rightarrow (1,0)$.

Sometimes it is useful to study the set of all positive limitpoints of an orbit $\gamma$; this set is called the $\omega$-limitset of $\gamma$. The set of all negative limitpoints of $\gamma$ is called the $\alpha$-limitset. These sets are indicated with $\omega(\gamma)$ and $\alpha(\gamma)$.
In example 4.4 the $\omega$-limitset of all orbits is the origin; the $\alpha$-limitset is empty for all orbits except when starting in the origin.
In example 4.5 we have for each orbit $\gamma$:

$$\omega(\gamma) = \alpha(\gamma) = \gamma.$$

We shall formulate and discuss some properties of limitsets.

**Theorem 4.2**
The sets $\alpha(\gamma)$ and $\omega(\gamma)$ are closed and invariant. Furthermore we have that if the

positive orbit $\gamma^+$ is bounded, the $\omega$-limitset is compact, connected and not empty. If $x(t; x_0)$ corresponds with $\gamma^+(x_0)$ for $t \geq 0$, we have that the distance $d(x(t; x_0), \omega(\gamma)) \to 0$ for $t \to \infty$ (the analogous property holds for the $\alpha$-limitset if $\gamma^-$ is bounded).

**Proof**

It follows from the definition that $\alpha(\gamma)$ and $\omega(\gamma)$ are closed. We shall prove first that $\omega(\gamma)$ is positive invariant. Suppose the limitpoint $p$ is contained in $\omega(\gamma)$, so there is a sequence $t_n \to \infty$ as $n \to \infty$ so that $x(t; x_0)$, which corresponds with $\gamma(x_0)$, tends to $p$ as $n \to \infty$ for these values of $t$. It follows from the translation property in lemma 2.1 that

$$x(t + t_n; x_0) = x(t; x(t_n; x_0)).$$

Taking the limit for $n \to \infty$ with $t$ fixed we have

$$x(t + t_n; x_0) \to x(t; p).$$

We conclude that the orbit which contains $p$ lies in $\omega(\gamma)$, so $\omega(\gamma)$ is positive invariant.

If $\gamma^+$ is bounded, $\omega(\gamma)$ is bounded and not empty as each bounded set in $\mathbf{R}^n$ with an infinite number of points has at least one point of accumulation. We noted already that $\omega(\gamma)$ is closed so $\omega(\gamma)$ is compact. We have proved that for each $t$ $x(t + t_n; x_0) \to x(t, p)$ where $\gamma(p) \subset \omega(\gamma(x_0))$. It follows that

$$d(x(t; x_0), \omega(\gamma)) \to 0 \text{ as } t \to \infty,$$

so $\omega(\gamma)$ is connected.
Analogous reasoning holds for $\omega(\gamma)$. □

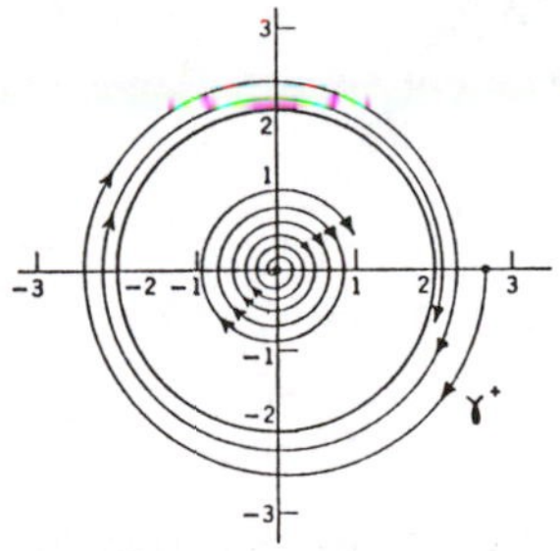

*Figure 4.3*

Theorem 4.2 does not seem to be very concrete but the theorem has important practical consequences. Obviously the limitsets $\alpha(\gamma)$ and $\omega(\gamma)$ do contain complete orbits only (if $p \in \omega(\gamma)$ than $x(t; p) \subset \omega(\gamma)$). We have seen this in the examples 4.4 and 4.5. At the same time this gives insight in the way an orbit $\gamma$, which does not correspond with a periodic solution, can approach a closed orbit. The orbit $\gamma$ cannot end up in one point of the closed orbit as this has to be a solution, in this case a critical point. This, however, contradicts with the assumption that we have one closed orbit. The orbit $\gamma$ approaches the closed orbit arbitrarily close but will keep on moving around the closed orbit (figure 4.3). This property will be illustrated again quantitatively in the theory of stability, chapters 5 and 7. We conclude this section with another auxiliary concept and a theorem.

**Definition**
A set $M \subset \mathbb{R}^n$ is called a minimal set of equation 4.3 $\dot{x} = f(x)$ if $M$ is closed, invariant, not empty and if $M$ has no smaller subsets with these three properties.

**Theorem 4.3**
Suppose that $A$ is a nonempty, compact (bounded and closed), invariant set of equation 4.3 ($\dot{x} = f(x)$ in $\mathbb{R}^n$) then there exists a minimal set $M \subset A$.

**Proof**
See Hale (1969), section 1.8. □

In the examples 4.4 and 4.5 the $\omega$-limitsets are all minimal (the reader should check this). We shall consider another two examples of autonomous systems in $\mathbb{R}^2$ where polar coordinates, $r$, $\theta$ are introduced by

$$x = r\cos\theta \, , \, y = r\sin\theta.$$

**Example 4.6**
$\dot{y} = y + x - y(x^2 + y^2)$, $\dot{x} = x - y - x(x^2 + y^2)$ which transforms to $\dot{r} = r(1 - r^2)$ , $\dot{\theta} = 1$.

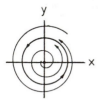

*Figure 4.4*

All circular domains with centre (0,0) and radius larger than 1 are positive invariant. Moreover, (0,0) and the circle $\{r = 1\}$ are invariant sets. The origin is $\alpha$-limitset for each orbit which starts with $r(0) < 1$. The circle $\{r = 1\}$, corresponding with the periodic solution $(x_1, x_2) = (\cos t, \sin t)$ is $\omega$-limitset for all orbits except the critical point in the origin. Both the $\alpha$- and the $\omega$-limitset are minimal.

**Example 4.7**
(Hale (1969), section 1.8).

$$\dot{r} = r(1-r)$$
$$\dot{\theta} = \sin^2\theta + (1-r)^3.$$

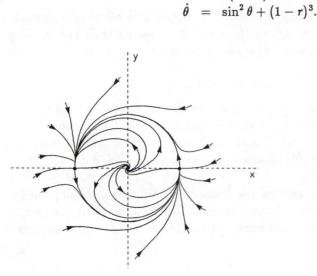

*Figure 4.5*

The origin and the circle $\{r = 1\}$ are invariant sets. The circle $\{r = 1\}$ is also $\omega$-limitset for all orbits starting outside the origin and outside the circle. The invariant set $\{r = 1\}$ consists of 4 orbits, given by $\theta = 0$ , $\theta = \pi$ and the arcs $0 < \theta < \pi$, $\pi < \theta < 2\pi$. The $\omega$-limitset $\{r = 1\}$ has two minimal sets, the points $\{r = 1, \theta = 0\}$ and $\{r = 1, \theta = \pi\}$.

It will be clear that, to understand the general behaviour of solutions of autonomous equations of the form 4.3, a classification of the possible minimal sets is essential. We have the beginning of a classification for dimension $n = 2$, for $n \geq 3$ this classification is very difficult and incomplete.

## 4.3 The Poincaré-Bendixson theorem

We shall restrict ourselves now to planar systems. Consider the equation

(4.4) $$\dot{x} = f(x)$$

with $x \in \mathbb{R}^2$; $f : \mathbb{R}^2 \to \mathbb{R}^2$ has continuous first partial derivatives and we assume that the solutions we are considering, exist for $-\infty < t + \infty$.

The Poincaré-Bendixson theorem consists of the nice result that, having a positive orbit $\gamma^+$ of equation 4.4, which is bounded but does not correspond with a periodic solution, the $\omega$-limitset $\omega(\gamma^+)$ contains a critical point or it consists of a closed orbit.

In the proof the following topological property is essential: a sectionally smooth, closed orbit $C$ in $\mathbf{R}^2$, a cycle, separates the plane into two parts $S_e$ and $S_i$. The sets $S_e$ and $S_i$ are disjunct and open, $C$ is the boundary of both $S_e$ and $S_i$; we can write

$$\mathbf{R}^2/C = S_e \cup S_i \quad \text{(Jordan)}.$$

Two simple tools will be useful. First we shall talk of phase points which are not critical points of equation 4.4; such a point we shall call an *ordinary point*.

Secondly the concept of *transversal*. The smooth curve $l$ will be called a transversal of the orbits of equation 4.4 if $l$ contains ordinary points only and if $l$ is nowhere tangent to an orbit; see figure 4.6

There exist several, related proofs for the Poincaré-Bendixson theorem; see for instance Coddington and Levinson (1955). Here we are using basically the formulation given by Hale (1969). We shall construct the proof out of several lemmas.

*Figure 4.6*

Consider the transversal $l$ which is supposed to be a closed set, the interior of $l$ is $l_0$. We consider now in particular the set $V \subset l_0$ which consists of the points p, mapped by the planar phase flow into $l_0$. So we assume that there exists a time $t_{p_1}$ such that the solution which starts in $p_1 \in l_0 : x(t; p_1)$ has the property that $x(t; p_1)$ is contained in $\mathbf{R}^2/l$ for $0 < t < t_{p_1}$, while $x(t_{p_1}; p_1) = p_2 \in l_0$.

This mapping of $V$ into $l_0$ is called a Poincaré-mapping. We shall analyse this mapping of $V$ into $l_0$ by the construction of a homeomorphism (one-to-one mapping which is continuous in both directions) $h : [-1, 1] \to l$ with $W = h^{-1}(V)$. The mapping $g : W \to (-1, 1)$ arises by using $h$ to map $w$ on $p_1 \in l_0$, $p_1$ is mapped by the Poincaré-mapping (phase flow) on $p_2$, $p_2$ is mapped in its turn on $g(w)$ by $h^{-1}$; altogether we have $g(w) = h^{-1}\phi(t_{h(w)}, h(w))$. The properties of $g$ tell us something about the Poincaré-mapping $V$ into $l_0$, in particular about the monotonical character of this mapping.

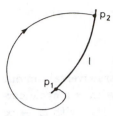

*Figure 4.7*

## Lemma 4.1

The set $W$ is open; the function $g(w)$ is continuous and non-decreasing in $W$; the sequence $\{g^k(w)\}_{k=0}^n$ is monotonic ($g^0(w) = w$, $g^1(w) = g(g^0(w))$ etc.).

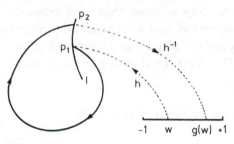

*Figure 4.8*

## Proof

Using the implicit function theorem, we note that an open neighbourhood of $p_1 = h(w)$ is mapped on an open neighbourhood of $p_2$ in $l_0$, so $W$ is open. The continuity follows from the fact that the solutions $\phi(t_{h(w)}, h(w))$ depend continuously on the initial values $h(w)$.

Consider now the curve $C$ which consists of the orbit connecting $p_1$ and $p_2$ and the curve $p_1 p_2$ lying in $l_0$. If $p_1 = p_2$ then $\gamma(p_1)$ corresponds with a periodic orbit; $p_1$ is in that case a fixed point of the mapping of $V$ into $l_0$. The sequence $\{g^k\}$ consists of one point. Now suppose $p_1 \neq p_2$ and for instance $h^{-1}(p_1) < h^{-1}(p_2)$. $C_i$ is the interior part of $C$, $C_e$ the exterior part. As the transversal $l$ is nowhere tangent to the orbits, $l$ can be intersected in one direction only. It follows that $h[g(w), 1] \subset l_0$ must be contained in $C_e$. If $g^2(w)$ is defined, this point must be in $(g(w), +1)$. Induction yields the monotonicity of the sequence.

## Remark

The reader will find it useful to give a more detailed presentation of the assertion in the first sequence of the proof.

Using lemma 4.1 and the various mappings involved, will give us insight in the possible $\omega$-limitsets in $\mathbf{R}^2$.

## Lemma 4.2

The $\omega$-limitset of an orbit $\gamma(p)$ can intersect the interior $l_0$ of a transversal $l$ in one point only. If there exists such a point $p_0$, we have $\omega(\gamma) = \gamma$ with $\gamma$ a cycle corresponding with a periodic solution, or there exists a sequence $\{t_k\}$ with $t_k \to \infty$ such that $\phi(t_k, p)$ tends to $p_0$ monotonically in $l_0$.

## Proof

Suppose that $\omega(\gamma)$ intersects $l_0$ in $p_0$. If $p = p_0$, the orbit is periodic. If not, $\omega(\gamma)$ being the limitset of $\gamma(p)$ means that there exists a sequence $\{t_k^1\}$, $t_k^1 \to \infty$ as $k \to \infty$, with $x(t_k^1, p) \to p_0$ as $k \to \infty$. Then, there exists a sequence $\{t_k\}$, $t_k \to \infty$ as $k \to \infty$, in a neighbourhood of $p_0$ such that $x(t_k, p) \to p_0$ in $l_0$. From lemma 4.1 we have that this sequence is monotonic in $l_0$. Now suppose that $\omega(\gamma)$ intersects in $l_0$ in the point $p_0^1$; with the same reasoning we can find a sequence $x(\tau_k, p)$ in $l_0$ which tends to $p_0^1$. So we can construct a sequence $\{g^k(\omega)\}$ in the sense of lemma 4.1 which is not monotonic unless $p_0 = p_0^1$. The existence of such a sequence which is not monotonic is prohibited by lemma 4.1. $\qquad\square$

It is now possible to prove an important result for minimal sets in $I\!R^2$.

## Theorem 4.4

If $M$ is a bounded, minimal set of equation 4.4 $\dot{x} = f(x)$ in $\mathbf{R}^2$, then $M$ is a critical point or a periodic orbit.

## Proof

The set $M$ is not empty, bounded and invariant and so contains at least one orbit $\gamma$. The limitsets $\alpha(\gamma)$ and $\omega(\gamma)$ are also contained in $M$.
If $M$ contains a critical point, the minimal character of $M$ means that $M$ has to be identified with this critical point.
If $M$ contains no critical points, $\gamma$ and $\omega(\gamma)$ consist of ordinary points. Because of the minimal character of $M$ we have that $\gamma \subset \omega(\gamma)$. Choose a point $p \in \gamma$ and a transversal $l_0$ of $\gamma$ in $p$. It follows from lemma 4.2 that either $\gamma = \omega(\gamma)$ and $\gamma$ is periodic or there exists a sequence in $l_0$ which tends monotonically to $p$. The latter alternative is prohibited by lemma 4.1 so $\gamma$ is periodic. $\qquad\square$

At this point we still need one lemma.

## Lemma 4.3

If $\omega(\gamma^+)$ contains both ordinary points and a periodic orbit $\gamma_0$ then we have $\omega(\gamma^+) = \gamma_0$.

**Proof**

Suppose that $\omega(\gamma^+)\backslash\gamma_0$ is not empty.

Choose $p_0 \in \gamma_0$ and an open transversal $l_0$ which contains $p_0$. It follows from lemma 4.2 that $\omega(\gamma^+)\cap l_0 = \{p_0\}$. Moreover there exists a sequence $\{p_n\}$ in $\omega(\gamma^+)\backslash\gamma_0$ such that $p_n \to p_0$ as $n \to \infty$. For $n$ sufficiently large $\gamma(p_n)$ will intersect the transversal $l_0$. According to lemma 4.2 however, this can happen in one point only. This is a contradiction. □

The theorems which we have proved lead to the following result.

**Theorem 4.5** (Poincaré-Bendixson)

Consider equation 4.4 $\dot{x} = f(x)$ in $\mathbf{R}^2$ and assume that $\gamma^+$ is a bounded, positive orbit and that $\omega(\gamma^+)$ contains ordinary points only. Then $\omega(\gamma^+)$ is a periodic orbit. If $\omega(\gamma^+) \neq \gamma^+$ the periodic orbit is called a limit cycle. An analogous result is valid for a bounded, negative orbit.

**Proof**

It follows from theorem 4.2 that $\omega(\gamma^+)$ is compact, connected and not empty. Then theorem 4.3 yields that there exists a bounded, minimal set $M \subset \omega(\gamma^+)$; $M$ contains ordinary points only. Because of theorem 4.4 $M$ is a periodic orbit which has, using lemma 4.3, the properties formulated in the theorem.

The proof of the Poincaré-Bendixson theorem is very simple: we have needed five lines only.

## 4.4   Applications of the Poincaré-Bendixson theorem

To apply the theorem one has to find a domain $D$ in $\mathbf{R}^2$ which contains ordinary points only and one has to find at least one orbit which for $t \geq 0$ enters the domain $D$ without leaving it. Then $D$ must contain at least one periodic orbit.

The prototype example is again example 4.6 in section 4.2. There we considered the system

$$\dot{y} = y + x - y(x^2 + y^2)$$
$$\dot{x} = x - y - x(x^2 + y^2).$$

The only critical point is (0,0); this is a spiral point with negative attraction. We construct an annular domain with centre in (0,0) and inner radius $r_1 < 1$ and outer radius $r_2 > 1$. Because of the negative attraction of the origin, the orbits

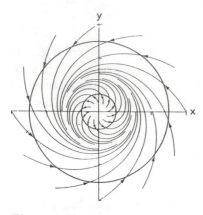

*Figure 4.9*

which are starting inside the smallest circle will enter the annulus. One can analyse the isoclines of the system to conclude that orbits which are starting outside
the largest circle will also enter the annulus. As the annulus contains no critical
points, according to the Poincaré-Bendixson theorem the annulus must contain at
least one periodic orbit.

This analysis is confirmed, in a rather trivial way, by transforming to polar coordinates:

$$\dot{r} = r(1 - r^2) \, , \, \dot{\theta} = 1.$$

There exists one periodic solution: $r(t) = 1$, $\theta(t) = \theta_0 + t$. In the phase plane this
is a limit cycle.

Somewhat less trivial is the following problem.

## Example 4.8
Consider the system

$$\dot{y} = y(x^2 + y^2 - 2x - 3) + x$$
$$\dot{x} = x(x^2 + y^2 - 2x - 3) - y$$

The only critical point is (0,0); this is a spiral point with positive attraction. To
see whether closed orbits are possible we apply the criterion of Bendixson, theorem
4.1. We find for the divergence of the vector function on the righthand-side

$$4x^2 + 4y^2 - 6x - 6 = 4[(x - \frac{3}{4})^2 + y^2 - \frac{33}{16}].$$

Inside the (Bendixson-) circle with centre $(\frac{3}{4}, 0)$ and radius $\sqrt{33}/4$ the expression
is sign definite and no closed orbits can be contained in the interior of this circle.

Closed orbits are possible which enclose or which intersect this Bendixson-circle.

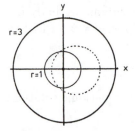

*Figure 4.10 ··· Bendixson circle*

We transform the system to polar coordinates by $x = r \cos \theta$, $y = r \sin \theta$ to find

$$\begin{aligned}\dot{r} &= r(r^2 - 2r \cos \theta - 3) \\ \dot{\theta} &= 1.\end{aligned}$$

If $r < 0$ we have $\dot{r} < 0$, if $r > 3$ we have $\dot{r} > 0$. According to the Poincaré-Bendixson theorem the annulus $1 < r < 3$ must contain one or more limit cycles. For a more detailed analysis we have to use quantitative, for instance numerical, methods which are not discussed here.

**Example 4.9**(the equations of Liénard and van der Pol)
Consider the equation of Liénard

(4.5) $$\ddot{x} + f(x)\dot{x} + x = 0$$

with $f(x)$ Lipschitz-continuous in R. We assume that
a. $F(x) = \int_0^x f(s)ds$ is an odd function.
b. $F(x) \to +\infty$ as $x \to \infty$ and there exists a constant $\beta > 0$ such that for $x > \beta$, $F(x) > 0$ and monotonically increasing.
c. There exists a constant $\alpha > 0$ such that for $0 < x < \alpha$, $F(x) < 0$.

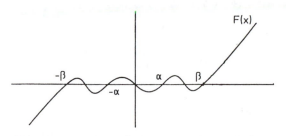

*Figure 4.11*

In the figure the typical behaviour of $F(x)$ has been sketched where the number of oscillations of $F(x)$ between $-\beta$ and $\beta$ has been chosen at random. In analysing equation 4.5 it turns out to be convenient to transform $x, \dot{x} \to x, y(= \dot{x} + F(x))$ so that

(4.6)
$$\begin{aligned} \dot{x} &= y - F(x) \\ \dot{y} &= -x \end{aligned}$$

We shall show that equation 4.5 (or system 4.6) has at least one periodic solution:

## Theorem 4.6

Consider the equation of Liénard 4.5 or the equivalent system 4.6. If the conditions $a - c$ have been satisfied, the equation has at least one periodic solution. If moreover $\alpha = \beta$, there exists only one periodic solution and the corresponding orbit is $\omega$-limitset for all orbits except the critical point $(0,0)$.

## Proof

First we remark that system 4.6 has only one critical point: $(0,0)$. By expansion in a neighbourhood of $(0,0)$ we find

$$\begin{aligned} \dot{x} &= y - F'(0)x + \cdots \\ \dot{y} &= -x. \end{aligned}$$

With $F'(0) = f(0)$ the eigenvalues are

$$\lambda_{1,2} = -\frac{1}{2}f(0) \pm \frac{1}{2}[f^2(0) - 4]^{\frac{1}{2}}.$$

It follows from the assumptions that $F'(0)$ is negative, so $(0,0)$ is a negative attractor.

To apply the Poincaré-Bendixson theorem we prove also for this system, that there exists an annular domain which is positive invariant. We shall use again the polar coordinate $r$, or to simplify the expressions

$$R = \frac{1}{2}r^2 = \frac{1}{2}(x^2 + y^2).$$

For the solutions of system 4.6 we find

$$\dot{R} = x\dot{x} + y\dot{y} = x(y - F(x)) + y(-x)$$

or

$$\dot{R} = -xF(x).$$

We note that for $-\alpha < x < \alpha$ we have $\dot{R} \geq 0$; this is in agreement with the negative attraction of $(0,0)$. Orbits starting on the boundary of a circular domain with radius smaller than $\alpha$, cannot enter this circular domain. We shall show now, that if orbits are starting far away from the origin, the orbits will decrease the distance from the origin. First we remark that on replacing $(x, y)$ in system 4.6 by $(-x, -y)$ the system does not change as $F(x)$ is odd. This means that if $(x(t), y(t))$ is a solution, reflection through the origin yields again a solution: $(-x(t), -y(t))$. We investigate the behaviour of the orbit which starts in $(0, y_0)$ with $y_0 > 0$; from system 4.6 we find the equation for the orbits

$$(4.7) \qquad \frac{dy}{dx} = -\frac{x}{y - F(x)}.$$

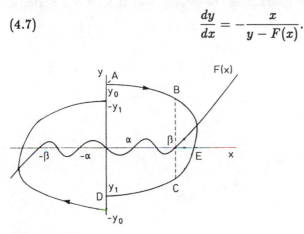

Figure 4.12

The tangent to the orbit is horizontal if $x = 0$, vertical if $y = F(x)$. We shall show that on choosing $y_0$ large enough, the behaviour of the solution starting in $(0, y_0)$ is like the behaviour sketched in figure 4.12 where $|y_1| < y_0$. If this is the case, the proof is complete as reflection of the orbit produces an invariant set, bounded by the two orbits and the segments $[-y_1, y_0]$, $[-y_0, y_1]$.

To prove that $|y_1| < y_0$ we consider $R(x, y)$, in particular

$$R(0, y_1) - R(0, y_0) = \int_{ABECD} dR =$$

$$= (\int_{AB} + \int_{CD}) \frac{-xF(x)}{y - F(x)} dx + \int_{BEC} dR.$$

$F(x)$ is bounded for $0 \le x \le \beta$ so the first expression tends to zero as $y_0 \to \infty$; note that we assumed that $y_1 \to -\infty$ as $y_0 \to \infty$, if not, the proof would be finished.

Using equation 4.7 we write the second expression in a different way.

$$\int_{BEC} dR = \int_{BEC} F(x) dy.$$

If $x > \beta$ we have $F(x) > 0$; the integration is carried out from positive values of $y$ to negative values so the integral is negative. The integral approaches $-\infty$

as $y_0 \to \infty$ because of the unbounded increase of length of the curve $BEC$. We conclude that if $y_0$ is large enough

$$R(0, y_1) - R(0, y_0) = \int_{ABECD} dR < 0.$$

The Poincaré-Bendixson theorem guarantees the existence of at least one periodic solution.

The case $\alpha = \beta$.

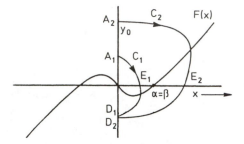

*Figure 4.13*

If one chooses $y_0$ sufficiently small, the orbits behave like $c_1$. We estimate $R_{D_1} - R_{A_1} = \int_{A_1 E_1 D_1} F(x) dy > 0$ as $F(x) < 0$ for $0 < x < \alpha$. So no periodic orbit can start in $(x_0, 0)$ with $0 < x_0 < \alpha$.

Now consider a curve $C_2$ intersecting the $x$-axis in $E_2$, to the right of $(\alpha, 0)$.

$$I = R_{D_2} - R_{A_2} = \int_{A_2 E_2 D_2} F(x) dy.$$

For $x \geq \alpha$, $F(x)$ is monotonically increasing from 0 to $+\infty$. We know already that this integral tends to $-\infty$ as $y_0 \to \infty$. Because of the monotonicity of $F(x)$, the integral $I$ has one zero, i.e. one $y_0$ such that $R_{A_2} = R_{D_2}$ and so one periodic solution. $\square$

This proof is typical for the 'calculus with hands and feet' which one uses for a nontrivial application of the Poincaré-Bendixson theorem. The theorem is based on profound ideas, the applications are often complicated but elementary.

Another remark is that theorem 4.6 provides us with the proof of the existence of a unique periodic solution of the van der Pol equation (cf. section 2.3)

$$\ddot{x} + x = \mu(1 - x^2)\dot{x} , \mu > 0.$$

In this case we have $f(x) = \mu(x^2 - 1)$, $F(x) = \mu(\frac{1}{3}x^3 - x)$. The conditions $a$, $b$ and $c$ of theorem 4.6 have been satisfied; $\alpha = \beta = \sqrt{3}$.

Other applications of the Poincaré-Bendixson theorem are concerned with generalisations of the Liénard equation

$$\ddot{x} + f(x)\dot{x} + g(x) = 0.$$

Assuming certain properties of $g(x)$ we can give an analogous proof of theorem 4.6 for this equation. See for instance the books by Sansone and Conti (1964) and Cesari (1971).

**Remark** on the number of limit cycles of plane system.
An extensive literature exists for polynomial equations of the form

$$\dot{x} = P(x,y) \ , \ \dot{y} = Q(x,y)$$

with $P$ and $Q$ polynomials in $x$ and $y$. The van der Pol-equation is a special case but there are many other applications. A survey for equations in which the polynomials are quadratic has been presented by Reyn (1987); the paper contains many phase portraits referring to various cases.
A classical problem in this respect has been posed by Hilbert as part of his so-called "sixteenth problem". The problem is that, when given the set of polynomial second-order equations of degree $n$, one wants to put an upperbound on the number of possible limit cycles. This turns out to be a very difficult and, as yet, unsolved problem. For quadratic systems it has been shown that there cannot be infinitely many limit cycles and examples have been constructed with four limit cycles. For a survey of methods and literature the reader is referred to Ye Yan-Qian (1986), where important work of Chinese mathematicians has been discussed, and Lloyd (1987).

## 4.5 Periodic solutions in $\mathbf{R}^n$

There are many results in the literature for periodic solutions of systems with dimension larger than two, but the theory is far from complete. A summary of results and methods can be found in the books of Sansone and Conti (1964) and Cesari (1971) regarding the older literature. Apart from these, one can consult the book by Hale (1969), especially for applications of the implicit function theorem, and the book by Amann (1983), where the concept of degree of a mapping and Brouwer's fixed point theorem have been used.
We note that the theory of periodic solutions will be discussed again in the sections 10.4 and 11.6, the Poincaré continuation method and the method of averaging, which are tied in with the implicit function theorem. Also periodic solutions play an important part in the theory of Hamiltonian systems, chapter 15.

Modern research in the theory of periodic solutions uses functional analytic and topological methods. To illustrate the ideas we discuss some results.

In section 2.3 we saw that because of the translation property of solutions of autonomous equations, periodic solutions of these equations correspond with closed orbits in phase space. Considering, more generally, non-autonomous equations it is useful to study the so-called $T$-mapping $a_T$.

Consider the equation $\dot{x} = f(t, x)$ in $\mathbb{R}^n$. The solution $x(t; x_0)$ starts at $t = t_0$ in $x_0$; such a point $x_0$ yields the point $x(t_0 + T; x_0)$ with $T$ a constant, chosen such that the solution exists for $0 \leq t - t_0 \leq T$. The set $D \subset \mathbb{R}^n$ of points $x_0$ to which we can assign such a point $x(t_0 + T; x_0)$ is mapped by this $T$-mapping $a_T$ into $\mathbb{R}^n$: $a_T \colon D \to \mathbb{R}^n$. The mapping $a_T$ is parameterised by $T$, $t_0$ and $x_0$. The point $x_0$ is a fixed point of $a_T$ if $a_T(t_0, x_0) = x_0$ or $x(t_0 + T; x_0) = x_0$.

**Example 4.10**

Consider in $\mathbb{R}^n$ the system

$$\dot{x} = 2x + \sin t$$
$$\dot{y} = -y$$

The solutions are known explicitly so it is easy to construct the mapping $a_T(t_0, x_0, y_0)$. For the set $D$ we choose for instance a square in the first quadrant, see figure 4.14.

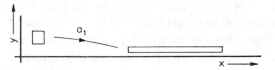

Figure 4.14

The mapping $a_1(0, x_0, y_0)$ produces a rectangle in the first quadrant. Applying the mapping $a_{2\pi}(0, x_0, y_0)$ the picture is qualitatively the same, quantitatively of course not. There are no fixed points in these two cases. Next we choose for $D$ a square with side length 1, centred in $(0,0)$. The mapping $a_1(0, x_0, y_0)$ produces a contraction in the $y$-direction and an asymmetric expansion in the $x$-direction. There are no fixed points, see figure 4.15.

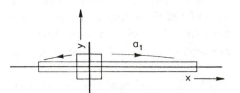

Figure 4.15

Finally we apply the mapping $a_{2\pi}(0, x_0.y_0)$ to the set $D$ of figure 4.16. The result looks roughly the same, but there is a qualitative difference as a fixed point $(-\frac{1}{5}, 0)$ exists. For reasons of comparison we give the solutions starting for $t = 0$

$$
\begin{aligned}
x(t) &= (x_0 + \tfrac{1}{5})e^{2t} - \tfrac{2}{5}\sin t - \tfrac{1}{5}\cos t \\
y(t) &= y_0 e^{-t}
\end{aligned}
$$

There exists clearly one periodic solution which starts in $(-\frac{1}{5}, 0)$, period $2\pi$.

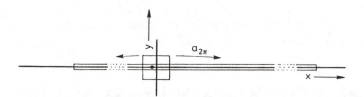

*Figure 4.16*

In a number of cases a fixed point of the mapping $a_T$ corresponds with a $T$- periodic solution of the equation.

**Lemma 4.4**
Consider in $\mathbf{R}^n$ the equation $\dot{x} = f(t, x)$ with a righthand side which is $T$-periodic $f(t + T, x) = f(t, x)$, $t \in \mathbf{R}$, $x \in \mathbf{R}^n$. The equation has a $T$-periodic solution if and only if the $T$-mapping $a_T$ has a fixed point.

**Proof**
It is clear that a $T$-periodic solution produces a fixed point of $a_T$. Suppose now that $a_T$ has a fixed point $x_0$. Then we have $x(t_0 + T; x_0) = x_0$ for a certain solution $x(t; x_0)$ starting at $t = t_0$. The vector function $x(t + T; x_0)$ will also be a solution as

$$
\dot{x} = f(t, x) = f(t + T, x).
$$

As $x_0$ is a fixed point of $a_T$, $x(t + T; x_0)$ has the same initial value $x_0$ as $x(t; x_0)$. Because of uniqueness we then have $x(t + T; x_0) = x(t; x_0)$, $t \in \mathbf{R}$, so $x(t; x_0)$ is $T$-periodic. □

The difficulty in applying these ideas is of course to show the existence of a fixed point without explicit knowledge of the solutions. In some cases one can apply the following fixed point theorem.

**Theorem 4.7 (Brouwer)**
Consider a compact, convex set $V \subset \mathbf{R}^n$ which is not empty. Each continuous mapping of $V$ into itself has at least one fixed point.

**Proof**
The proof is difficult; see for instance Amann (1983). □

This fixed point theorem was generalized by Schauder for Banach spaces; other extensions are associated with the names of Tychonov and Browder. In the sections 4.3 and 4.4 we met positive invariant sets, annular regions, containing a periodic solution but no critical point. In the following application of Brouwer's fixed point theorem we shall show that a positive invariant set, which is convex, contains at least one critical point.

## Theorem 4.8
Consider the equation $\dot{x} = f(x)$, $x \in \mathbb{R}^n$ with positive invariant, compact, convex set $V \subset \mathbb{R}^n$. The equation has at least one critical point in $V$.

## Remark
In section 2.3 we mentioned that $T$-periodic solutions will have a fixed period $T > 0$ unless explicitly stated. Here we have such a case. In the proof of theorem 4.8 a critical point will be interpreted as a periodic solution with arbitrary period $T \geq 0$.

## Proof
Consider for each $T > 0$ the $T$-mapping $a_T : V \to V$. It follows from Brouwer's theorem that for any $T > 0$ the mapping $a_T$ has at least one fixed point and so the equation contains a $T$-periodic solution by lemma 4.4. We choose a sequence $T_1, T_2, \cdots, T_m, \cdots \to 0$ as $m \to \infty$ with corresponding fixed points $p_m$ of $a_{T_m}$: $x(T_m; p_m) = p_m$. A subsequence of fixed points converges towards the fixed point $p$ and we indicate the points of this subsequence again by $p_m, m = 1, 2, \cdots$ with periods $T_m$. We estimate

$$\|x(t; p) - p\| \leq \|x(t; p) - x(t; p_m)\| + \|x(t; p_m) - p_m\| +$$

$$+ \|p_m - p\|$$

where we have used the triangle inequality twice. Consider the righthand side. The last term becomes arbitrary small as $m \to \infty$ because of the convergence of the subsequence; the first term becomes arbitrary small as $m \to \infty$ because of this convergence and continuity. The second term is estimated as follows; $x(t; p_m)$ is $T_m$-periodic, so for any $t > 0$ we can find $N \in I\!N$ such that

$$\begin{aligned} x(t; p_m) &= x(NT_m + qT_m; p_m), 0 \leq q < 1 \\ &= x(qT_m; p_m). \end{aligned}$$

By definition we have $p_m = x(0; p_m)$. Also $T_m \to 0$ as $m \to \infty$ so that $\|x(t; p_m) - p_m\| \to 0$ for any $t$. So $x(t; p) = p$ for any $t$. $\qquad\square$

In a number of cases one can prove the existence of a $T$-periodic solution of certain $T$-periodic equations by using Brouwer's theorem (4.7) and lemma 4.4. This is important as it is often very difficult to obtain this knowledge without explicit constructions of solutions.

The qualitative methods of this section have, on the other hand, rather serious limitations: in most cases these methods provide us with a lower bound only of the number of periodic solutions; this can be a lower bound far removed from the actual number of periodic solutions. These theorems, moreover, do not lead us to a localisation of the periodic solutions so that we still have to carry out explicit calculations.

To conclude and also to illustrate this discussion, we give another example of a theorem which has been based on the Brouwer fixed point theorem.

**Theorem 4.9**

Consider the equation $\dot{x} = A(t)x + g(t, x)$, $t \in R, x \in R^n$. The matrix $A(t)$ is continuous in $t$, the vector function $g(t, x)$ is continuous in $t$ and $x$, Lipschitz-continuous in $x$; $A(t)$ and $g(t, x)$ are $T$-periodic in $t$. Moreover

$$g(t, x) = o(\|x\|) \text{ as } \|x\| \to \infty$$

uniformly in $t$ (the terms in the equation are dominated by the linear part as $\|x\| \to \infty$). Finally we assume that the equation $\dot{y} = A(t)y$ has no $T$-periodic solutions except the solution $y = \dot{y} = 0$.

With these assumptions, the equation for $x$ has at least one $T$-periodic solution.

**Proof**

See Amann (1983), chapter 5, section 22. □

**Example 4.11**

Consider the scalar equation

$$\ddot{x} + \omega^2 x = \frac{1 + x}{1 + x^2} \cos t.$$

Theorem 4.9 guarantees the existence of at least one $2\pi$-periodic solution if $\omega^2 \neq 1$. Interesting questions are: where is this periodic solution located, are there more periodic solutions, what is the behaviour of a periodic solution if one varies $\omega$, what happens if $\omega = 1$?

Other applications of Brouwer's theorem are concerned with the forced Liénard equation

$$\ddot{x} + f(x)\dot{x} + g(x) = F(t)$$

with $F(t)$ $T$-periodic. For references and a discussion see Cesari (1971), chapter 9.6.

## 4.6 Exercises

4-1. In exercise 2.3 of chapter 2 we analysed the existence of periodic solutions in an invariant set of a three-dimensional system. Obtain this result in a more straightforward manner.

4-2. Consider the Liénard equation

$$\ddot{x} + f(x)\dot{x} + x = 0$$

with $f(x) = a + bx + cx^2 + dx^3$.
Determine the parameters $a \ldots d$ such that the phase-plane contains limit cycles. How many are there?

4-3. In the system $\dot{x} = f(y), \dot{y} = g(x) + y^k$ are $f$ and $g$ $C^1$ functions, $k \in \mathbb{N}$.

   a. Give sufficient conditions for $k$ so that the system contains no periodic solutions.

   b. Choose $f(y) = -y, g(x) = x, k = 2$. Does the system contain periodic solutions, cycles or limit cycles?

4-4. An equation arising in applications has been called after Rayleigh

$$\ddot{x} + x = \mu(1 - \dot{x}^2)\dot{x} \ , \mu > 0.$$

Show that the equation has a unique periodic solution by relating it to the van der Pol-equation.

4-5. Consider again the system from excercise 3.1

$$\dot{x} = y(1 + x - y^2)$$
$$\dot{y} = x(1 + y - x^2)$$

but suppose that the equations model an experimental situation such that $x \geq 0, y \geq 0$ (for instance because $x$ and $y$ are quantities in chemical reactions). Do periodic solutions exist?

4-6. Consider the system
$$\dot{x} = y - x^3 + \mu x$$
$$\dot{y} = -x$$

   a. For which values of the parameter $\mu$ does a periodic solution exist?

   b. Describe what happens as $\mu \downarrow 0$.

4-7. In $\mathbf{R}^n$ we consider the equation $\dot{x} = f(x)$ and a point $a$; $f(x)$ is continuously differentiable. Suppose that a solution $\varphi(t)$ of the equation exists such that

$$\lim_{t \to \infty} \varphi(t) = a.$$

Show that $a$ is a critical point of the equation.

4-8. In this exercise we show that the $w$-limitset of an orbit of a plane system is not necessarily a critical point or a closed orbit. Consider

$$\begin{aligned} \dot{x} &= \frac{\partial E}{\partial y} + \lambda E \frac{\partial E}{\partial x} \\ \dot{y} &= -\frac{\partial E}{\partial x} + \lambda E \frac{\partial E}{\partial y} \end{aligned}$$

with $\lambda \in \mathbf{R}$, $E(x,y) = y^2 - 2x^2 + x^4$.

a. Put $\lambda = 0$. Determine the critical points and their character by linear analysis. What happens in the nonlinear system; sketch the phase-plane.

b. What happens to the critical points of $a$ if $\lambda \neq 0$.

c. Choose $\lambda < 0$; the orbit $\gamma^+$ starts at $t = 0$ in $(\frac{1}{2}, 0)$, $\gamma_2^+$ in $(-\frac{1}{2}, 0)$, $\gamma_3^+$ in $(1, 2)$. Determine the $w$-limit sets of these orbits.

# 5   Introduction to the theory of stability

## 5.1   Simple examples

In the chapters three and four we have seen equilibrium solutions and periodic solutions. These are solutions which exist for all time. In applications one is often interested also in the question whether solutions which at $t = t_0$ are starting in a neighbourhood of such a special solution, will stay in this neighbourhood for $t > t_0$. If this is the case, the special solution is called stable and one expects that this solution can be realised in the practice of the field of application: a small perturbation does not cause the solutions to move away from this special solution. In mathematics these ideas pose difficult questions. In defining stability, this concept turns out to have many aspects. Also there is of course the problem that in investigating the stability of a special solution, one has to characterise the behaviour of a set of solutions. One solution is often difficult enough.

**Example 5.1**

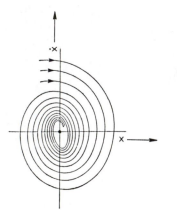

*Figure 5.1*

Consider the harmonic oscillator with damping

$$\ddot{x} + \mu\dot{x} + x = 0 \ , \ \mu > 0.$$

The critical point $x = \dot{x} = 0$ corresponds with an equilibrium solution. From the behaviour of the solutions and of the orbits in the phase plane we conclude, that $(0,0)$ is a positive attractor. It seems natural to call $(0,0)$ stable.

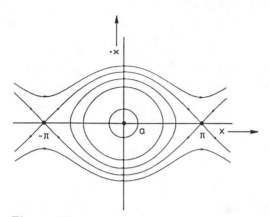

Figure 5.2

## Example 5.2

Consider again the mathematical pendulum, described by the equation

$$\ddot{x} + \sin x = 0.$$

The equilibrium solutions $(\pi, 0)$ and $(-\pi, 0)$ are called unstable as solutions starting in a neighbourhood of $(\pi, 0)$ or $(-\pi, 0)$ will generally leave this neighbourhood. Solutions in a neighbourhood of the equilibrium point $(0,0)$ behave differently from the solutions of example 5.1 near $(0,0)$. Still, also in this example it seems natural to call $(0,0)$ stable.

Consider now a periodic solution starting in $x(0) = a$, $\dot{x}(0) = 0$ with $0 < a < \pi$. Solutions starting in a neighbourhood are also periodic and they are staying nearby. Is the periodic solution starting in $(a, 0)$ stable?

To answer this question, we have to decide first how to define stability. Note that solutions starting in a neighbourhood of $(a, 0)$ have various periods. This means that phase points starting near each other are not necessarily staying close. We consider this in more detail.

The equation has the integral $\frac{1}{2}\dot{x}^2 - \cos x = -\cos a$ (compare for instance example 2.11 in section 2.4). It follows that

$$\frac{dx}{dt} = \pm(2\cos x - 2\cos a)^{\frac{1}{2}}.$$

Because of the symmetry of the periodic solutions we have for the period

$$T = 4 \int_0^a \frac{dx}{(2\cos x - 2\cos a)^{\frac{1}{2}}}.$$

It is clear that the period $T$ depends non-trivially on $x(0) = a$; see figure 5.3. It will become clear in the subsequent sections that the solutions in the examples 5.1-2, which we labeled stable in a intuitive way, are indeed stable but in different meanings of the word.

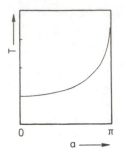

0           $\pi$

$a \longrightarrow$

*Figure 5.3*

## 5.2 Stability of equilibrium solutions

We are considering the equation

(5.1) $$\dot{x} = f(t,x) \ , \ x \in \mathbf{R}^n, t \in \mathbf{R}$$

with $f(t,x)$ continuous in $t$ and $x$, Lipschitz-continuous in $x$. Assume that $x = 0$ is a critical point of the vector function $f(t,x)$, so $f(t,0) = 0$, $t \in \mathbf{R}$. Critical points of non-autonomous equations are fairly rare, often the equation will be autonomous. The assumption $x = 0$ for the critical point is of course no restriction as we can translate any critical point to the origin of phase space; cf. section 2.2.

**Definition** (stability in the sense of Lyapunov)
Consider the equation 5.1 and a neighbourhood $D \subset \mathbf{R}^n$ of $x = 0$; the solutions starting at $t = t_0$ in $x = x_0 \in D$ are indicated by $x(t; t_0, x_0)$. The solution $x = 0$ is called stable in the sense of Lyapunov (or Lyapunov-stable) if for each $\epsilon > 0$ and $t_0$ a $\delta(\epsilon, t_0) > 0$ can be found such that $\|x_0\| \leq \delta$ yields $\|x(t; t_0, x_0)\| \leq \epsilon$ for $t \geq t_0$.

It is easy to see that the equilibrium solution $(0,0)$ of the equation for the mathematical pendulum (example 5.2 in section 5.1) is stable in the sense of Lyapunov; the same result holds for the equilibrium solution of the harmonic oscillator and of the harmonic oscillator with damping (example 5.1).
The property does not hold for the solutions $(\pm\pi, 0)$ of the equation for the mathematical pendulum. Such solutions are called unstable.

**Definition**
Consider the equilibrium solution $x = 0$ of equation 5.1. If this solution is not stable in the sense of Lyapunov, it is called unstable.

In the case of the harmonic oscillator with damping, the solutions do not only stay in a neigbhourhood of the equilibrium solution, they approach it with time. This is an example of a stronger type of stability:

**Definition**

The equilibrium solution $x = 0$ of equation 5.1 is called asymptotically stable if $x = 0$ is stable and if there exists a $\delta(t_0) > 0$ such that

$$\|x_0\| \leq \delta(t_0) \Rightarrow \lim_{t \to \infty} \|x(t; t_0, x_0)\| = 0.$$

The examples of linear systems in the sections 2.2 and 3.1-2 with $x = 0$ a positive attractor, are examples of equations with asymptotically stable solution $x = 0$. However, it should be noted that on the other hand, positive attraction is not sufficient for asymptotic stability. We shall illustrate this as follows.

**Example 5.3**

Consider the two-dimensional system which in polar coordinates ($x = r\cos\theta, y = r\sin\theta$) looks like

$$\begin{aligned} \dot{r} &= r(1-r) \\ \dot{\theta} &= \sin^2(\theta/2). \end{aligned}$$

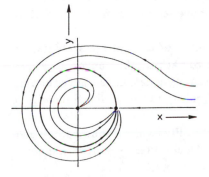

*Figure 5.4*

There are two critical points, $(r, \theta) = (0,0)$ and $(1,0)$. The point $(1,0)$ is positive attractor ($\omega$-limitset) for each orbit which starts outside $(0,0)$. However, both the solutions $(0,0)$ and $(1,0)$ are unstable. For in each neighbourhood of $(1,0)$ one can find solutions which are leaving this neighbourhood, although only temporarily. This is a disquieting example, but it should directly be made clear that the example is rather pathological; more examples demonstrating unusual behaviour can be found in Bhatia and Szegö (1970). In chapter 7 we shall see, that for critical points which are non-degenerate (cf. section 3.1) this behaviour cannot occur: a critical point which, after linearisation, corresponds with a positive attractor, turns out to be asymptotically stable.

## 5.3 Stability of periodic solutions

Consider again equation 5.1

$$\dot{x} = f(t, x)$$

which is assumed to satisfy the conditions of the theorem of uniqueness and existence of solutions. It is a straightforward affair to extend the definitions of Lyapunov-stability in section 5.2.

**Definition** (Stability in the sense of Lyapunov for periodic solutions)
Consider equation 5.1 with periodic solution $\phi(t)$. The periodic solution is Lyapunov-stable if for each $t_0$ and $\epsilon > 0$ we can find $\delta(\epsilon, t_0) > 0$ such that

$$\|x_0 - \phi(t_0)\| \leq \delta \Rightarrow \|x(t; t_0, x_0) - \phi(t)\| \leq \epsilon \text{ for } t \geq t_0.$$

We remark immediately that a periodic solution being Lyapunov-stable will be an exceptional case. For this implies that orbits, starting in a neighbourhood of the periodic solution, remain in a neighbourhood, but in the sense that phase points which start out being close, stay near each other. This has for example been realised for the harmonic oscillator equation, the solutions of which are synchronous (all solutions have the same period). For the periodic solutions of the mathematical pendulum equation this is not the case, as we have seen in example 5.2, section 5.1. So these solutions are not Lyapunov-stable.

More complications arise if one turns to the concept of asymptotic stability. In the discussion of theorem 4.2 in section 4.2 we remarked, that an orbit $\gamma$ in a neighbourhood of a periodic solution of an autonomous equation can never end up in a point of the periodic orbit. The orbit $\gamma$ continues to move around the periodic orbit.

These considerations will lead to definitions of stability, which require some geometric ideas. In the subsequent section we shall discuss linear perturbation equations of periodic solutions.

First consider the autonomous equation in $\mathbf{R}^n$

$$(5.2) \qquad \qquad \dot{x} = f(x)$$

with periodic solution $\phi(t)$, corresponding with a closed orbit in $n$- dimensional phase-space. Now, we construct a $(n-1)$-dimensional transversal $V$ to the closed orbit. With the transversal $V$ we denote a manifold, punctured by the closed orbit and nowhere tangent to it.

The closed orbit intersects the transversal in the point $a$. Consider now an orbit $\gamma(x_0)$ starting in $x_0 \in V$; we follow the orbit until it returns in $V$. If this is not the case, we choose $x_0$ nearer to $a$. Because of continuous dependence of the solution on the initial value, we may conclude that the phase-flow in a certain neighbourhood of the closed orbit will return to $V$.

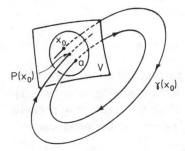

*Figure 5.5*

Suppose the point $x_0 \in V$ is mapped into $V$ by the phase-flow. This mapping is called the return-map or Poincaré-map $P$. The point $P(x_0) \in V$ which we have found, can be mapped again: $P^2(x_0) \in V$ etc. The point $a \in V$ is a fixed point of $P$. In section 4.3 we already saw an example of a Poincaré-map in a two-dimensional phase-space; the transversal is then one-dimensional.

The concept of stability of the periodic solution shall now be re-phrased as stability of the Poincaré-map in the neighbourhood of the fixed point $a$.

**Definition**
Given is the autonomous equation 5.2 with periodic solution $\phi(t)$, transversal $V$ and Poincaré-map $P$ with fixed point $a$. The solution $\phi(t)$ is stable if for each $\epsilon > 0$ we can find $\delta(\epsilon)$ such that

$$\|x_0 - a\| \leq \delta, x_0 \in V \Rightarrow \|P^n(x_0) - a\| \leq \epsilon\, , n = 1, 2, 3, \cdots$$

**Remark**
In some books, a periodic solution which is stable in this sense, is called orbitally stable.

It will be clear that the periodic solutions of the mathematical pendulum equation are stable. Consider the equation $\ddot{x} + \sin x = 0$ with periodic solution starting in $x(0) = a$, $\dot{x}(0) = 0$, $0 < a < \pi$. Take for $V$ a segment $I \subset (0, \pi)$ which has $a$ as interior point. The corresponding Poincaré-map is the identity mapping.

The concept of asymptotic stability of a periodic solution shall now be connected with the attraction properties of the fixed point $a$, when applying the Poincaré-map in a neighbourhood.

**Definition**
Given is the autonomous equation 5.2 with periodic solution $\phi(t)$, transversal $V$ and Poincaré-map $P$ with fixed point $a$. The solution $\phi(t)$ is asymptotically stable

if it is stable and if there exists a $\delta > 0$ such that

$$\|x_0 - a\| \leq \delta, x_0 \in V \Rightarrow \lim_{n \to \infty} P^n(x_0) = a.$$

We have introduced the concepts of stability and of asymptotic stability by using a mapping, which is for the equation $\dot{x} = f(x)$ in $\mathbf{R}^n$ a mapping of a $(n-1)$-dimensional transversal manifold into itself. How shall we treat now periodic solutions of non-autonomous equations? Consider in particular the equation

(5.3) $$\dot{x} = f(t, x) , x \in \mathbf{R}^n, t \in \mathbf{R},$$

$f(t, x)$ is $T$-periodic in $t$. Equation 5.3 is equivalent with the $(n+1)$- dimensional autonomous system

(5.4) $$\begin{aligned} \dot{x} &= f(\theta, x) \\ \dot{\theta} &= 1, \theta(0) = 0. \end{aligned}$$

with $(\theta, x) \in S^1 \times \mathbf{R}^n$ ($S^1$ is the circle $\mathbf{R} \bmod T$).
We can apply the previous definitions to system 5.4; a $T$-periodic solution $(\theta, x)$ of 5.4 corresponds with a $T$-periodic solution $x$ of 5.3. The transversal $V$ is now $n$-dimensional; a natural choice is to use the mapping $\mathbf{R}^n \to \mathbf{R}^n$ which arises by taking the value of solutions at time $T$, $2T$, etc. We clarify this by an example.

**Example 5.4** (linear oscillations with damping and forcing)
Consider the equation
(5.5) $$\ddot{x} + 2\mu\dot{x} + \omega_0^2 x = h \cos \omega t$$

with $0 < \mu < \omega_0, \mu$ denotes the damping rate; $h$ and $\omega$ are positive, $h$ is the amplitude of the periodic forcing; $\omega_0$ is the (positive) frequency of the free ($\mu = h = 0$) oscillating system. Without restriction of generality we may choose $\omega_0 = 1$ (or transform $\omega_0 t = \tau$) but to see the influence of the parameters it is more transparent to leave them as they are.
Written in the form of system 5.4, equation 5.5 becomes

(5.6) $$\begin{aligned} \dot{x}_1 &= x_2 \\ \dot{x}_2 &= -\omega_0^2 x_1 - 2\mu x_2 + h \cos \omega \theta \\ \dot{\theta} &= 1, \theta(0) = 0, \end{aligned}$$

The solutions of equation 5.5 are of the form

(5.7) $$x(t) = c_1 e^{-\mu t} \cos \sqrt{\omega_0^2 - \mu^2} t + c_2 e^{-\mu t} \sin \sqrt{\omega_0^2 - \mu^2} t + \alpha \cos \omega t$$

$$+ \beta \sin \omega t$$

where
$$\alpha = \frac{\omega_0^2 - \omega^2}{4\mu^2\omega^2 + (\omega_0^2 - \omega^2)^2} h , \beta = \frac{2\mu\omega}{4\mu^2\omega^2 + (\omega_0^2 - \omega^2)^2} h.$$

We have excluded the case of resonance $\mu = 0$, $\omega = \omega_0$. The constants $c_1$ and $c_2$ are determined by the initial values $x(0)$ and $\dot{x}(0)$. Substitution into 5.7 yields

$$c_1 = x(0) - \alpha$$
$$c_2 = [\dot{x}(0) + \mu x(0) - \mu\alpha - \omega\beta]/\sqrt{\omega_0^2 - \mu^2}.$$

Equation 5.5 (and system 5.6) has one periodic solution, period $2\pi/\omega$; the initial values are determined by $c_1 = c_2 = 0$. It follows that for this periodic solution

$$x(0) = \alpha, \ \dot{x}(0) = \omega\beta.$$

We shall construct now a Poincaré-mapping $P$ by considering at time $t = 0$, $2\pi/\omega$, $4\pi/\omega, \cdots$ the intersection of the orbit with the $x_1, x_2$-plane. The point $(\alpha, \omega\beta)$ is fixed point of $P$. In this case, where we have the solutions in terms of elementary functions, it is possible to give the mapping $P$ explicitly.

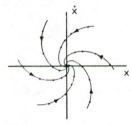

Figure 5.6

Substitution of $t = 2\pi/\omega$ produces with $\gamma = \sqrt{\omega_0^2 - \mu^2}\,\frac{2\pi}{\omega}$

$$P\left(\begin{array}{c} x(0) \\ \dot{x}(0) \end{array}\right) = P\left(\begin{array}{c} x_1(0) \\ x_2(0) \end{array}\right) =$$

$$\left(\begin{array}{c} c_1 e^{-\mu 2\pi/\omega}\cos\gamma + c_2 e^{-\mu 2\pi/\omega}\sin\gamma + \alpha \\ e^{-\mu 2\pi/\omega}\cos\gamma(-c_1\mu + c_2\sqrt{\omega_0^2 - \mu^2} - e^{-\mu 2\pi/\omega}\sin\gamma(c_1\sqrt{\omega_0^2 - \mu^2} + \mu c_2) + \omega\beta \end{array}\right).$$

Applying $P$ again and again we find

$$\lim_{n\to\infty} P^n\left(\begin{array}{c} x(0) \\ \dot{x}(0) \end{array}\right) = \left(\begin{array}{c} \alpha \\ \omega\beta \end{array}\right).$$

which means that the periodic solution is asymptotically stable. In this example it is of course simpler to draw this conclusion directly from the expression 5.7. In figure 5.6 we have indicated $P^n(x(0), \dot{x}(0))$ for various values of $n$ and several initial values.

## 5.4 Linearisation

In section 2.2 we saw how to obtain linear quations by linearisation of the vector function in a neighbourhood of a critical point. For non-autonomous equations we have the same procedure. Suppose that $x = a$ is a critical point of equation 5.1

$$\dot{x} = f(t, x).$$

In a neighbourhood of $x = a$ we consider the linear equation

(5.8) $$\dot{y} = \frac{\partial f}{\partial x}(t, a)y.$$

We shall now give a similar formulation for periodic solutions. Suppose that $\phi(t)$ is a periodic solution of equation 5.1. We put

$$x = \phi(t) + y$$

and after substitution and expansion we obtain

$$\dot{\phi}(t) + \dot{y} = f(t, \phi(t) + y)$$
$$= f(t, \phi(t)) + \frac{\partial f}{\partial x}(t, \phi(t))y + \cdots$$

We have assumed here that $f(t, x)$ has a Taylor expansion to degree two. As $\phi(t)$ satisfies the equation we have

(5.9) $$\dot{y} = \frac{\partial f}{\partial x}(t, \phi(t))y + \cdots$$

Linearisation means that we omit the higher order terms $(\cdots)$.

The equation which arises by linearisation of 5.9 is in general difficult to solve. If $f(t, x)$ and $\phi(t)$ are both $T$-periodic, with as a special case that $f(t, x)$ does not depend explicitly on $t$, the linearised equation has coefficients which are $T$-periodic. In this case the so-called theory of Flocquet applies which will be described in the next chapter.

The equation obtained by linearisation of 5.9 has $n$ independent solutions. If $f(t, x)$ does not depend explicitly on $t$, the equation is autonomous,

$$\dot{x} = f(x)$$

and we immediately know one of the solutions of the linearised equation. Suppose that $\phi(t)$ is a $T$-periodic solution, so

$$\dot{\phi}(t) = f(\phi(t))$$

and

$$\ddot{\phi}(t) = \frac{\partial f}{\partial x}(\phi(t))\dot{\phi}(t).$$

So the derivative $\dot{\phi}(t)$ satisfies the linearised equation! One of the implications is that, if the dimension of the vectorfunction $f(x)$ is 2, we can solve the linearised equation by explicitly constructing the second independent solution (see Walter, 1976 or any elementary textbook on ordinary differential equations).

Apart from linearisation of the equation in a neighbourhood of a periodic solution, one can study the linearisation of the Poincaré-mapping $P$ in the corresponding fixed point $x = a$. This means that for the mapping $P$ of $x_0$ we linearise in a neighbourhood of $x = a$ so that we have to calculate

$$\frac{\partial P}{\partial x_0}(a)x_0.$$

In the case of the autonomous equation $\dot{x} = f(x)$ in $\mathbf{R}^n$, the square matrix of coefficients has dimension $n - 1$. In the example of section 5.3 with non-autonomous terms (forced oscillation) the matrix has dimension 2.

In the subsequent chapter 6, we shall summarise the information on linear equations which we are needing. In chapter 7 we shall discuss the important question of the relation between the nonlinear and the linearised equation.

## 5.5  Exercises

5-1. Determine the stability properties of the following solutions:

    a. $(\cos t, \sin t)$ of $\begin{array}{rcl} \dot{x} &=& -y(x^2 + y^2)^{-1/2} \\ \dot{y} &=& x(x^2 + y^2)^{-1/2} \end{array}$

    b. $(0,0)$ of $\ddot{x} + \alpha\dot{x} + x = 0$, $\alpha \in \mathbf{R}$.

    c. the periodic solution of $\begin{array}{l} \dot{x} = x - y - x(x^2 + y^2) \\ \dot{y} = x + y - y(x^2 + y^2). \end{array}$

5-2. Consider the system
$$\dot{x} = y$$
$$\dot{y} = f(x,y) - x \text{ with } \begin{array}{rcl} f(x,y) &=& y\sin^2\left(\frac{\pi}{x^2+y^2}\right), (x,y) \neq (0,0) \\ f(0,0) &=& 0 \end{array}$$
Determine the stability of $(0,0)$

5-3. In example 5.2 we discussed the nontrivial dependence of the period of the mathematical pendulum on the initial value $(x, \dot{x}) = (a, 0)$. Demonstrate this dependence more explicitly for small values of $a$.

5-4. Consider the second-order equation

$$\ddot{x} + x^2 + bx + c = 0.$$

    a. Determine the critical points and their stability.

    b. Give a condition for periodic solutions to exist. Are they in this case Lyapunov-stable, stable?

5-5. Determine the stability of the solution $(0,0)$ of

$$
\begin{aligned}
\dot{x} &= -ax + y + ay^2, a \in \mathbb{R} \\
\dot{y} &= (1-a)x + xy.
\end{aligned}
$$

# 6  Linear equations

There is an abundance of theorems for linear equations but still there are many difficult and unsolved problems left. This chapter contains a summary of a number of important results.

## 6.1  Equations with constant coefficients

Consider the equation

(6.1)
$$\dot{x} = Ax$$

with $A$ a non-singular, constant $n \times n$-matrix. The eigenvalues $\lambda_1, \cdots, \lambda_n$ are solutions of the characteristic equation

(6.2)
$$\det(A - \lambda I) = 0.$$

If $n \geq 4$, the calculation of the solutions of the characteristic equation can amount already to a lot of work.

Suppose that the eigenvalues $\lambda_k$ are distinct with corresponding eigenvectors $c_k$, $k = 1, \cdots, n$. In this case

$$c_k e^{\lambda_k t} , \ k = 1, \cdots, n$$

are $n$ independent solutions of equation 6.1.

Suppose now that not all eigenvalues are distinct, for instance the eigenvalue $\lambda$ has multiplicity $m > 1$. This eigenvalue $\lambda$ generates $m$ independent solutions of the form

$$P_0 e^{\lambda t}, P_1(t) e^{\lambda t}, P_2(t) e^{\lambda t}, \cdots, P_{m-1}(t) e^{\lambda t}$$

where $P_k(t)$ , $k = 0, 1, \cdots, m - 1$ are polynomial vectors of degree $k$ or smaller. For the development of the theory it is useful to compose $n$ independent solutions $x_1(t), \cdots, x_n(t)$ of equation 6.1 to a matrix $\Phi(t)$ with these solutions as columns:

$$\Phi(t) = (x_1(t) x_2(t) \ldots x_n(t)).$$

$\Phi(t)$ is called a fundamental matrix of equation 6.1. Each solution of equation 6.1 can be written as

$$x(t) = \Phi(t)c$$

with $c$ a constant vector. Adding the initial value condition $x(t_0) = x_0$ to equation 6.1 we have as solution of the initial value problem

(6.3)
$$x(t) = \Phi(t)\Phi^{-1}(t_0)x_0.$$

Often one chooses the fundamental matrix $\Phi(t)$ such that $\Phi(t_0) = I$, the $n \times n$ identity matrix.

When studying the stability of the solution $x = 0$, it is clear this is mainly determined by the real part of the eigenvalues. From 6.3 and the explicit form of the independent solutions we have immediately the following result:

**Theorem 6.1**

Consider equation 6.1 , $\dot{x} = Ax$, with $A$ a non-singular, constant $n \times n$-matrix, eigenvalues $\lambda_1, \cdots, \lambda_n$.

a. If $Re\lambda_k < 0$, $k = 1, \cdots, n$, then for each $x(t_0) = x_0 \in \mathbb{R}^n$ and suitably chosen positive constants $C$ and $\mu$ we have

$$\|x(t)\| \leq C\|x_0\|e^{-\mu t} \text{ and } \lim_{t \to \infty} x(t) = 0.$$

b. If $Re\lambda_k \leq 0, k = 1, \cdots, n$, where the eigenvalues with $Re\lambda_k = 0$ are distinct, then $x(t)$ is bounded for $t \geq t_0$. Explicitly

$$\|x(t)\| \leq C\|x_0\|$$

with $C$ a positive constant.

c. If there exists an eigenvalue $\lambda_k$ with $Re\lambda_k > 0$, then in each neighbourhood of $x = 0$ there are initial values such that for the corresponding solutions we have

$$\lim_{t \to \infty} \|x(t)\| = +\infty.$$

In the case $a$, the solution $x = 0$ is asymptotically stable, in the case $b$ $x = 0$ is Lyapunov-stable and in the case $c$ unstable.

**Remark 1**

There are a number of criteria to determine whether the roots of the characteristic equation have negative real parts or not. Well-known is the Routh-Hurwitz criterion; for details on these criteria the reader should consult Cesari (1971), section 2.4 and Hahn (1967), chapter 2.

**Example 6.1**

The equation $\dot{x} = Ax$ with

$$A = \begin{pmatrix} 1 & -a \\ 4 & -3 \end{pmatrix} \qquad , a \in \mathbb{R}$$

is studied which involves the calculation of the eigenvalues of the matrix $A$. We find for the eigenvalues $-1 \pm 2\sqrt{1-a}$ so the real part is negative if $a > \frac{3}{4}$. We have asymptotic stability of the trivial solution $x = 0$ from theorem 6.1a for $a > \frac{3}{4}$; the matrix $A$ is singular if $a = \frac{3}{4}$, the trivial solution is unstable if $a < \frac{3}{4}$. The reader will find that if $a > 1$, $x = 0$ is a spiral, for $\frac{3}{4} < a \leq 1$ we find a node, for $a < \frac{3}{4}$ the solution $x = 0$ represents a saddle.

**Remark 2**

The solutions of equation 6.1 $\dot{x} = Ax$ can be written in a different way, using the concept of exponential or exponential matrix

$$x(t) = e^{At}c.$$

The exponential function of a matrix is defined by the series expansion

$$e^{At} = \sum_{n=0}^{\infty} \frac{1}{n!}(At)^n = I + At + \frac{1}{2!}A^2 t^2 + \cdots$$

See for instance the introductory books by Walter (1976) and Arnold (1978). The concept of exponential is convenient for developing the theory. The fundamental matrix $\Phi(t)$ of equation 6.1 and its inverse can be written as

$$\Phi(t) = e^{At} \qquad , \Phi^{-1}(t) = e^{-At}.$$

## 6.2   Equations with coefficients which have a limit

Consider now the equation

(6.4) $$\dot{x} = Ax + B(t)x$$

with $A$ a non-singular, constant $n \times n$-matrix, $B(t)$ a continuous $n \times n$-matrix. An intuitive idea would be: if

$$\lim_{t \to \infty} \|B(t)\| = 0$$

then the solutions of equation 6.4 will tend to the solutions of equation 6.1

$$\dot{x} = Ax.$$

The idea to impose this condition on the matrix $B(t)$ is correct but in general not sufficient. We demonstrate this for a scalar equation.

**Example 6.2**

$$\ddot{x} - \frac{2}{t}\dot{x} + x = 0 \qquad , t \geq 1.$$

We might expect, that for $t \to \infty$ the solutions tend to the solutions of the equation

$$\ddot{x} + x = 0$$

which has bounded solutions only. However, the non-autonomous equation has the two independent solutions $\sin t - t \cos t$ and $\cos t + t \sin t$. These solutions have no upperbound if $t$ tends to infinity.

It turns out that, in the case of bounded solutions of the equation with constant coefficiënts 6.1, we have to require a little bit more of $B(t)$.

## Theorem 6.2
Consider equation 6.4 $\dot{x} = Ax + B(t)x$, $B(t)$ continuous for $t \geq t_0$ with the properties that

    a. the eigenvalues $\lambda_k$ of $A, k = 1, \cdots, n$ have $Re\lambda_k \leq 0$, the eigenvalues corresponding with $Re\lambda_k = 0$ are distinct;

    b. $\int_{t_0}^{\infty} \|B\| dt$ is bounded,

then the solutions of equation 6.4 are bounded and $x = 0$ is stable in the sense of Lyapunov.

## Proof
We shall use the method of 'variation of constants'.
The fundamental matrix $\Phi(t)$ of the equation $\dot{x} = Ax$ can be written as $\Phi(t) = \exp{(A(t - t_0))}$. We substitute $x = \Phi(t)z$ into equation 6.4 to obtain

$$\frac{d}{dt}\Phi(t)z + \Phi(t)\dot{z} = A\Phi(t)z + B(t)\Phi(t)z$$

and as $\frac{d}{dt}\Phi(t) = A\Phi(t)$ we find

$$\dot{z} = \Phi^{-1}(t)B(t)\Phi(t)z.$$

Integration of this expression and multiplication with $\Phi(t)$ produces for the solutions of equation 6.4 the integral equation

(6.5) $$x(t) = \Phi(t)x_0 + \int_{t_0}^{t} \Phi(t - \tau + t_0)B(\tau)x(\tau)d\tau.$$

Note that we have used that

$$\begin{aligned}\Phi(t)\Phi^{-1}(\tau) &= e^{A(t-t_0)}e^{-A(\tau-t_0)} \\ &= e^{A(t-\tau)} = \Phi(t - \tau + t_0).\end{aligned}$$

Equation 6.5 yields the inequality

(6.6) $$\|x\| \leq \|\Phi(t)\|\|x_0\| + \int_{t_0}^{t} \|\Phi(t - \tau + t_0)\|\|B(\tau)\|\|x(\tau)\|d\tau.$$

It follows from assumption $a$ and theorem 6.1 that

$$\|\Phi(t)\| \leq C \quad , t \geq t_0,$$

so the inequality 6.6 becomes

$$\|x\| \leq C\|x_0\| + \int_{t_0}^{t} C\|B(\tau)\|\|x\|d\tau.$$

Applying Gronwall's inequality, theorem 1.2 with $\delta_1 = 1$, $\delta_3 = C\|x_0\|$, we find

$$\|x\| \leq C\|x_0\| \exp. \left( C \int_{t_0}^{t} \|B(\tau)\|d\tau \right).$$

It follows from assumption $b$ that the righthand side and so $\|x\|$ is bounded, moreover it follows from the inequality that $x = 0$ is stable in the sense of Lyapunov. $\square$

If the real parts of the eigenvalues are all negative, a weaker assumption for the matrix $B(t)$ suffices.

**Theorem 6.3**
Consider equation 6.4 $\dot{x} = Ax + B(t)x$, $B(t)$ continuous for $t \geq t_0$ with

    a. $A$ is a constant matrix with eigenvalues $\lambda_k, k = 1, \cdots, n$ such that $Re\lambda_k < 0$;

    b. $\lim_{t \to \infty} \|B(t)\| = 0$

then for all solutions of equation 6.4 we have

$$\lim_{t \to \infty} x(t) = 0$$

and $x = 0$ is asymptotically stable.

**Proof**
As in the proof of theorem 6.2 we find inequality 6.6

$$\|x\| \leq \|\Phi(t)\|\|x_0\| + \int_{t_0}^{t} \|\Phi(t - \tau + t_0)\|\|B(\tau)\|\|x(\tau)\|d\tau.$$

We apply the estimate in theorem 6.1a for the fundamental matrix $\Phi(t)$

$$\|\Phi(t)\| \leq C_1 e^{-\mu(t-t_0)}$$

with suitable positive constants $C_1$ and $\mu$. It follows from assumption $b$, that for each $\varepsilon > 0$ there exists a time $t_1(\varepsilon) \geq t_0$ such that

$$\|B(t)\| \leq \varepsilon \quad , t \geq t_1(\varepsilon).$$

Both estimates will be used in inequality 6.6. for $t \geq t_1$:

(6.7)
$$\begin{aligned} \|x\| &\leq C_1 e^{-\mu(t-t_0)}\|x_0\| + \int_{t_0}^t C_1 e^{-\mu(t-\tau)}\|B(\tau)\|\|x(\tau)\|d\tau \\ &\leq C_1 e^{-\mu(t-t_0)}(\|x_0\| + \int_{t_0}^t e^{\mu(\tau-t_0)}\|B(\tau)\|\|x(\tau)\|d\tau + \\ & \quad \int_{t_1}^t \varepsilon e^{\mu(\tau-t_0)}\|x(\tau)\|d\tau). \end{aligned}$$

We now fix $\varepsilon > 0$ (and so $t_1(\varepsilon)$) such that

(6.8)
$$\varepsilon C_1 < \mu$$

Because of the linearity of equation 6.4 the solutions $x(t)$ exist for $t \geq t_0$ and so for $t_0 \leq t \leq t_1$. It follows that

$$\|x_0\| + \int_{t_0}^{t_1} e^{\mu(\tau-t_0)}\|B(\tau)\|\|x(\tau)\|d\tau \leq C_2$$

with $C_2$ a positive constant. We find with 6.7

$$e^{\mu(t-t_0)}\|x\| \leq C_1 C_2 + C_1 \varepsilon \int_{t_1}^t e^{\mu(\tau-t_0)}\|x(\tau)\|d\tau$$

and with Gronwall's inequality, theorem 1.2 with $\delta_1 = C_1\varepsilon$ and $\delta_3 = C_1 C_2$

$$e^{\mu(t-t_0)}\|x\| \leq C_1 C_2 \exp. ((C_1\varepsilon(t - t_1))$$

or

$$\|x\| \leq C_1 C_2 \exp. ((C_1\varepsilon - \mu)t + \mu t_0 - C_1 \varepsilon t_1).$$

Application of 6.8 completes the proof. □

## Example 6.3
To illustrate theorem 6.2 and 6.3 we consider some first order equations.
Consider $\dot{x} = -x + \frac{1}{1+t^2}x, t \geq 0$.
Theorem 6.2 applies, so $x = 0$ is stable; theorem 6.3 provides us with a stronger result: $x = 0$ is asymptotically stable.
Consider equation $\dot{x} = -x + \frac{1}{1+t}x, t \geq 0$. Theorem 6.2 does not apply, theorem 6.3 provides us with asymptotic stability of the solution $x = 0$.
Consider equation $\dot{x} = -x + \frac{at}{1+t}x, a > 0, t \geq 0$. The time-dependent coefficient is such that theorems 6.2 and 6.3 donot apply. By explicit integration we find asymptotic stability of $x = 0$ if $0 < a \leq 1$, instability if $a > 1$.

If the equation $\dot{x} = Ax + B(t)x$ has a matrix $A$ with some eigenvalues which have a positive, real part, then we expect the trivial solution to be unstable.

## Example 6.4
Consider the first order equation

$$\dot{x} = \lambda x + b(t)x, \quad \lambda > 0$$

where the continuous function $b(t)$ has the property that $\lim_{t\to\infty} b(t) = 0$. The solutions are given by

$$x(t) = x(t_0) \exp. \left(\lambda(t - t_0) + \int_{t_0}^{t} b(\tau)d\tau\right).$$

For some $t_1 \geq t_0$ we have $|b(t)| < \lambda$, $t \geq t_1$. So if $x(t_0) \neq 0$ we find

$$\lim_{t\to\infty} x(t) = +\infty.$$

In general the following theorem is useful.

**Theorem 6.4**
Consider equation 6.4 $\dot{x} = Ax + B(t)x$ with $B(t)$ continuous for $t \geq t_0$ and the property that $\lim_{t\to\infty} \|B(t)\| = 0$. If at least one eigenvalue of the matrix $A$ has a positive real part, there exist in each neighbourhood of $x = 0$ solutions $x(t)$ such that

$$\lim_{t\to\infty} \|x(t)\| = +\infty$$

The solution $x = 0$ is unstable.

**Proof**
See Roseau (1966), chapter 3. We shall not give the proof as it has much in common with the proof of theorem 7.3 in chapter 7; theorem 7.3 discusses equation 6.4 with nonlinearities added. □

To develop some feeling for the quantitative behaviour of solutions in a neighbourhood of an asymptotically stable equilibrium solution, we shall consider some elementary examples:

$$\begin{aligned}
\dot{x} &= -x \, , x(0) = 1 \\
\dot{y} &= (-1 + \tfrac{1}{1+t})y \, , y(0) = 1 \\
\dot{z} &= (-1 + 2e^{-t})z \, , z(0) = 1.
\end{aligned}$$

The reader should verify that the requirements of theorem 6.3 have been satisfied in these three cases: the trivial solution is asymptotically stable. Ths solutions are respectivily (see figure 6.1)

$$\begin{aligned}
x(t) &= e^{-t} \\
y(t) &= (1+t)e^{-t} \\
z(t) &= e^{(-t-2e^{-t}+2)}
\end{aligned} .$$

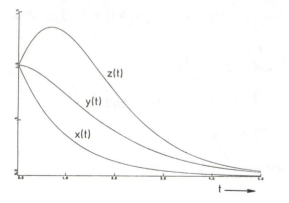

*Figure 6.1*

## 6.3 Equations with periodic coefficients

We shall consider the equation

(6.9)
$$\dot{x} = A(t)x \qquad , t \in \mathbf{R}$$

with $A(t)$ a continuous $T$-periodic $n \times n$-matrix; so $A(t+T) = A(t)$, $t \in \mathbf{R}$.
An equation like 6.9 can obtain both periodic and non-periodic solutions. Think for instance of the first order equation $\dot{x} = a(t)x$ with $a(t) = 1$, $t \in \mathbf{R}$ or $a(t) = \sin^2 t$. In both cases we have non-periodic, even unbounded solutions if $x(t_0) \neq 0$.
The Flocquet theorem contains the fundamental result for equations with periodic coefficients, that the fundamental matrix of equation 6.9 can be written as the product of a $T$-periodic matrix and a (generally) non-periodic matrix.

**Theorem 6.5 (Flocquet)**
Consider equation 6.9 $\dot{x} = A(t)x$ with $A(t)$ a continuous $T$-periodic $n \times n$-matrix. Each fundamental matrix $\Phi(t)$ of equation 6.9 can be written as the product of two $n \times n$-matrices
$$\Phi(t) = P(t)e^{Bt}$$
with $P(t)$ $T$-periodic and $B$ a constant $n \times n$-matrix.

**Proof**
The fundamental matrix $\Phi(t)$ is composed of $n$ independent solutions; $\Phi(t+T)$ is also a fundamental matrix. To show this, put $\tau = t + T$, then
$$\begin{aligned}\frac{dz}{d\tau} &= A(\tau - T)x \\ &= A(\tau)x\end{aligned}$$

So $\Phi(\tau)$ is also fundamental matrix. The fundamental matrices $\Phi(t)$ and $\Phi(\tau) = \Phi(t + T)$ are linearly dependent, which means that there exists a nonsingular

$n \times n$-matrix $C$ such that

$$\Phi(t+T) = \Phi(t)C.$$

There exists a constant matrix $B$ such that

$$C = e^{BT}$$

(if the reader is not familiar with these exponential formulations one should analyse the case of $C$ a diagonal matrix). We shall prove now that $\phi(t)$ exp. $(-Bt)$ is $T$-periodic. Put

$$\Phi(t)e^{-Bt} = P(t).$$

Then

$$\begin{aligned} P(t+T) &= \Phi(t+T)e^{-B(t+T)} \\ &= \Phi(t)Ce^{-BT}e^{-Bt} \\ &= \Phi(t)e^{-Bt} \\ &= P(t). \end{aligned}$$

$\square$

## Remark 1

The matrix $C$ which has been introduced is called the *monodromy*-matrix of equation 6.9. The eigenvalues $\rho$ of $C$ are called the *characteristic multipliers*. Each complex number $\lambda$ such that

$$\rho = e^{\lambda T}$$

is called a *characteristic exponent*. Sometimes the word 'characteristic' is replaced by 'Flocquet'. The imaginary parts of the characteristic exponents are not determined uniquely, we can add $2\pi i/T$ to them. The characteristic multipliers are determined uniquely. We can choose the exponents $\lambda$ such, that they coincide with the eigenvalues of the matrix $B$.

## Remark 2

The Flocquet theorem implies that the solutions of equation 6.9 consist of a product of polynomials in $t$, exp. $(\lambda t)$ and $T$-periodic terms. One way of using this result goes as follows. The equation

$$\dot{x} = A(t)x$$

can be transformed by

$$x = P(t)y$$

so that

$$\begin{aligned} \dot{P}(t)y + P(t)\dot{y} &= A(t)P(t)y \\ \text{or} \qquad\qquad \dot{y} &= P^{-1}(AP - \dot{P})y. \end{aligned}$$

On the other hand, differentiation of

$$P(t) = \Phi(t)e^{-Bt}$$

produces

$$\dot{P} = \dot{\Phi}e^{-Bt} + \Phi e^{-Bt}(-B)$$
$$= AP - PB.$$

So we find

$$\dot{y} = By.$$

In other words, the transformation $x = P(t)y$ carries equation 6.9 over into an equation with constant coefficients, the solutions of which are vector- polynomials in $t$ multiplied with exp.$(\lambda t)$.

This possibility of reduction of the linear part of the system to the case of constant coefficients will play a part in the theory of chapter 7.


## Remark 3

It will be clear that the existence of periodic solutions of equation 6.9 and the stability of the trivial solution are both determined by the eigenvalues of the matrix $B$. A necessary condition for the existence of $T$-periodic solutions is, that one or more of the characteristic exponents are purely imaginary (multiplier has modulus 1).

A necessary and sufficient condition for asymptotic stability of the trivial solution of equation 6.9 is, that all characteristic exponents have a negative real part (multipliers have modulus smaller than 1).

A necessary and sufficient condition for stability of the trivial solution is, that all characteristic exponents have real part $\leq 0$ while the exponents with real part zero have multiplicity one.


A serious problem of equations with periodic coefficients is, that there are no general methods available to calculate the matrix $P(t)$ or the characteristic exponents or multipliers. Each equation requires a special study and whole books have been devoted to some of them.

However, the following general theorem can be useful.


## Theorem 6.6

Equation 6.9 $\dot{x} = A(t)x$ has characteristic multipliers $\rho_i$ and exponents $\lambda_1, i = 1, \cdots, n$, $\rho_i = $ exp. $(\lambda_i T)$. Then we have the following expressions for the product of the multipliers and the sum of the exponents

$$\rho_1 \rho_2 \cdots \rho_n = \text{exp.} \left( \int_0^T Tr A(t)dt \right)$$
$$\sum_{i=1}^n \lambda_i = \tfrac{1}{T} \int_0^T Tr A(t)dt \, (\text{mod} \tfrac{2\pi i}{T}).$$

## Proof

Consider the fundamental matrix $\Phi(t)$ of equation 6.9 with $\Phi(0) = I$. For the determinant of $\Phi(t)$, the Wronskian, we have

$$\text{Det } \Phi(t) = \exp. \left( \int_0^t TrA(\tau)d\tau \right)$$

with $TrA(\tau)$, the trace of the matrix $A(\tau)$ (the sum of the diagonal elements). With theorem 6.5 we have

$$\text{Det } \Phi(t) = \text{Det. } (P(t)e^{Bt}).$$

Substituting $t = T$, equating the expressions for the Wronskian and using the $T$-periodicity of $P(t)$ we have

$$\text{Det. } (e^{BT}) = \exp. \left( \int_0^T TrA(t)dt \right).$$

Using the definitions of multipliers and exponents, we immediately find the required result. $\qquad \square$

It is clear that if on calculating the sum of the characteristic exponents, we find a positive number, the trivial solution is unstable. When finding a negative number or zero for this sum, we have not enough information to draw a conclusion about the stability of the trivial solution.

We have seen in section 5.4, when linearising in a neighbourhood of a periodic solution $\phi(t)$ of an autonomous equation, that one of the solutions of the linear system is $\dot\phi(t)$. This means, that in this case, if the equation has order 2, we can construct the other independent solution. For the characteristic exponents we have

$$\lambda_1 = 0 \qquad , \; \lambda_2 = \frac{1}{T} \int_0^T TrA(t)dt.$$

The calculation of $\lambda_2$ suffices if we are only interested in the problem of stability. We demonstrate this for the generalised Liénard equation of section 4.4.

**Application** (generalised Liénard equation)
Suppose, the equation

$$\ddot{x} + f(x)\dot{x} + g(x) = 0$$

has a $T$-periodic solution $x = \phi(t)$. Translating $x = \phi(t) + y$ and substitution produces

$$\ddot\phi + \ddot{y} + f(\phi + y)(\dot\phi + \dot{y}) + g(\phi + y) = 0$$

We expand $f$ and $g$ in a neighbourhood of $y = 0$ while indicating the nonlinear terms in $y$ and $\dot{y}$ by dots:

$$\ddot\phi + \ddot{y} + f(\phi)\dot\phi + \frac{df}{dx}(\phi)y\dot\phi + f(\phi)\dot{y} + \cdots + g(\phi) + \frac{dg}{dx}(\phi)y + \cdots = 0.$$

Three terms cancel as $\phi(t)$ satisfies the original equation. We are left with

$$\ddot{y} + f(\phi)\dot{y} + [\frac{df}{dx}(\phi)\dot{\phi} + \frac{dg}{dx}(\phi)]y = \cdots$$

Writing this equation in vector form by putting $y = y_1$, $\dot{y} = y_2$ we find

$$\begin{aligned} \dot{y}_1 &= y_2 \\ \dot{y}_2 &= -[\frac{df}{dx}(\phi)\dot{\phi} + \frac{dg}{dx}(\phi)]y_1 - f(\phi)y_2 + \cdots \end{aligned}$$

The trace of the linearized equation is

$$TrA(t) = -f(\phi(t)).$$

As we have seen in section 5.4 $\dot{\phi}(t)$ is a solution of the linearised equation. So we can put $\lambda_1 = 0$. Theorem 6.6 yields

$$\lambda_2 = -\frac{1}{T}\int_0^T f(\phi(t))dt \ (\mathrm{mod}\frac{2\pi i}{T}).$$

We find stability of the periodic solution in the linear approximation if $\lambda_2 \leq 0$, instability if $\lambda_2 > 0$.
One should note, that the stability of the periodic solution in the linear approximation is determined only by the term $f(x)$; the existence of the periodic solution is of course dependent also on $g(x)$.

It was mentioned earlier on, that the treatment of linear, periodic equations is not easy and the temptation is there to use heuristic methods. To discourage this for these problems we have the following example.

**Example 6.5 (Markus and Yamabe)**
Consider equation 6.9 $\dot{x} = A(t)x$ with $n = 2$ and

$$A(t) = \begin{pmatrix} -1 + \frac{3}{2}\cos^2 t & 1 - \frac{3}{2}\sin t\cos t \\ -1 - \frac{3}{2}\sin t\cos t & -1 + \frac{3}{2}\sin^2 t \end{pmatrix}.$$

Theorem 6.6 yields

$$\lambda_1 + \lambda_2 = \frac{1}{2\pi}\int_0^{2\pi}(-2 + \frac{3}{2}\cos^2 t + \frac{3}{2}\sin^2 t)dt = -\frac{1}{2}.$$

So we have no immediate conclusion about the stability of the trivial solution. As in the theory of equations with constant coefficients, one could calculate the time-dependent eigenvalues of $A(t)$ although it is not clear what the meaning is of such quantities. We find in this case for the eigenvalues $(-1 \pm i\sqrt{7})/4$, which surprisingly are time- independent. This suggests that the equation has characteristic exponents with negative real part and stability of the trivial solution. However, it can easily be checked, that a solution exists of the form

$$\begin{pmatrix} -\cos t \\ \sin t \end{pmatrix} e^{t/2}.$$

The characteristic exponents are $\lambda_1 = \frac{1}{2}, \lambda_2 = -1$, the trivial solution is unstable. Using this solution one can calculate a second independent solution; the fundamental matrix has the structure as given in the Flocquet theorem.

**Remark on the literature**
A lot of research has been put into Hill's equation

$$\ddot{x} + (a + b(t))x = 0, b(t + \pi) = b(t);$$

see for instance Magnus and Winkler (1966). A rich source of old and new results is the book by Jakubovič and Staržinskij (1975).
In appendix 2 we present some results on equations with small periodic coefficients.

## 6.4  Exercises

6-1. Consider the equation $\dot{x} = AX + B(t)x$ with $x \in \mathbf{R}^3$ and

$$A = \begin{pmatrix} -3 & 0 & 0 \\ 0 & -1 & 1 \\ 0 & 0 & -1 \end{pmatrix}, B(t) = \begin{pmatrix} e^{-t^2} & 0 & 0 \\ te^{-t^2} & t^2 e^{-t^2} & 0 \\ 0 & 0 & e^{-t^2} \end{pmatrix}.$$

Determine the stability of $x = 0$.

6-2. We are studying the equation $\dot{x} = A(t)x$ with $x \in \mathbf{R}^3$

$$A(t) = \begin{pmatrix} e^{-t} & \frac{t^2+1}{t^2} & e^{-t} \\ \frac{\sin t}{t^{3/2}} & 0 & 1 + e^{-t} \\ (2-a)(\frac{1-t}{t}) & -1 & a(\frac{1-t}{t}) \end{pmatrix}, t > 0, a \in \mathbf{R}.$$

For which values of $a$ is the solution $x = 0$ asymptotically stable or unstable (ignore the cases with real parts of eigenvalues zero).

6-3. $\dot{x} = Ax + B(t)x + C(t)$
with $x \in \mathbf{R}^n$, $A$ is a constant, non-singular $n \times n$-matrix, $B(t)$ is a continuous $n \times n$-matrix, $C(t)$ a continuous vector function in $\mathbf{R}^n$. Moreover we have:

- the eigenvalues $\lambda_i, i = 1, \ldots, n$ of the matrix $A$ have the property Re $(\lambda_i) \le 0, i = 1, \ldots, n$; eigenvalues with real part zero have multiplicity one;

- $\int_0^\infty \|B(t)\| dt = c_1$, $\int_0^\infty \|C(t)\| dt = c_2$ with $c_1$ and $c_2$ positive constants.

a. Prove that the solution $x(t)$ with $x(0) = x_0$ is bounded for all time.

b. Do we reach another conclusion on replacing $x(0) = x_0$ by $x(t_0) = x_0$?

c. Can we reach the result of $a$ with a weaker assumption on $C(t)$?

6-4. Prove the following theorem:
For the equation $\dot{x} = Ax + B(t)x$ we have $x \in \mathbb{R}^n$, $B(t)$ is continuous for $t \ge 0$, $A$ is a constant $n \times n$-matrix with eigenvalues $\lambda_k, k = 1, \ldots, n$ such that Re$\lambda_k < 0$. There exists a constant $b$ such that if $\|B(t)\| \le b$ for $t \ge 0$, it follows that $\lim_{t \to \infty} x(t) = 0$ and $x = 0$ is asymptotically stable.

6-5. The solution $(0,0)$ of the system

$$\begin{aligned} \dot{x} &= 2x + y + x\cos t - y\sin t \\ \dot{y} &= -x + 2y - x\cos t + y\sin t \end{aligned}$$

is unstable as $(e^{2t}\sin t, e^{2t}\cos t)$ is a solution of the system. Explain this. Do nontrivial solutions exist with the property $\lim_{t \to \infty} (x(t), y(t)) = 0$?

6-6. Consider the equation $\dot{x} = A(t)x$ with $x \in \mathbb{R}^2$ and

$$A = \begin{pmatrix} \frac{1}{2} - \cos t & b \\ a & \frac{3}{2} + \sin t \end{pmatrix}$$

and $a, b$ constants. Show that there exists at least a one-parameter family of solutions which becomes unbounded as $t \to \infty$.

6-7. Consider the equation $\dot{x} = A(t)x$ with $A(t)$ a smooth $T$-periodic $n \times n$-matrix, $x \in \mathbb{R}^n$, $f(t)$ a smooth scalar $T$- periodic function.

a. $n = 1, A(t) = f(t)$. Determine $P(t)$ and $B$ in the Flocquet theorem 6.5.

b. $n = 1, A(t) = f(t)$. Give necessary and sufficient conditions for the solutions to be bounded as $t \to \pm\infty$ or to be periodic.
Now we take $n = 2$ and

$$A(t) = f(t) \begin{pmatrix} a & b \\ c & d \end{pmatrix},$$

the matrix is constant.

c. Determine again $P(t)$ and $B$ in the Flocquet theorem 6.5.

d. Give necessary and sufficient conditions for the solutions to be bounded as $t \to \pm\infty$ or to be periodic.

e. Consider now the case $n = 2$ and

$$A(t) = \begin{pmatrix} \cos t & \sin t \\ \sin t & -\cos t \end{pmatrix}$$

Note that not only $Tr A(t) = 0$ but all terms of $A(t)$ have average zero. Are the solutions bounded?

# 7 Stability by linearisation

The stability of equilibrium solutions or of periodic solutions can be studied often by analysing the system, linearised in a neighbourhood of these special solutions. In section 5.4 we have discussed linearisation and we have given a summary of the analysis of linear systems. These methods have been in use for a long time but only since around 1900 the justification of linearisation methods has been started by Poincaré and Lyapunov.

A more recent result is the "stable and unstable manifold theorem" 3.3. This theorem is concerned with autonomous equations of the form

$$\dot{x} = Ax + g(x)$$

with $A$ a constant $n \times n$-matrix of which all eigenvalues have nonzero real part. Theorem 3.3 establishes that the stable and unstable manifolds $E_s$ and $E_u$ of the linearised equation $\dot{y} = Ay$ can be continued on adding the nonlinearity $g(x)$; the manifolds of the nonlinear equation $W_s$ and $W_u$, emanating from the origin, are tangent to $E_s$ and $E_u$. See again figure 3.15.

A natural question is what theorem 3.3 implies for the stability of the trivial solution. Note in this respect, that in the formulation of theorem 3.3 one only discusses the existence of invariant manifolds and not very explicitly the behaviour with time of individual solutions. In this chapter we shall add this quantitative element to the theory. Also, in the formulation of our theorems, we shall obtain more general results as we shall consider the stability of the trivial solution for nonautonomous equations.

## 7.1 Asymptotic stability of the trivial solution

**Theorem 7.1 (Poincaré-Lyapunov)**
Consider the equation in $\mathbb{R}^n$

$$(7.1) \qquad \dot{x} = Ax + B(t)x + f(t, x), x(t_0) = x_0 , t \in \mathbb{R}.$$

$A$ is a constant $n \times n$-matrix with eigenvalues which have all nonzero real part; $B(t)$ is a continuous $n \times n$-matrix with the property

$$\lim_{t \to \infty} \|B(t)\| = 0.$$

The vector function $f(t, x)$ is continuous in $t$ and $x$ and Lipschitz- continuous in $x$ in a neighbourhood of $x = 0$; moreover we have

$$\lim_{\|x\| \to 0} \frac{\|f(t, x)\|}{\|x\|} = 0 \text{ uniformly in } t$$

(this last condition also implies that $x = 0$ is a solution of equation 7.1).
Then there exist positive constants $C$, $t_0$, $\delta$, $\mu$ such that $\|x_0\| \leq \delta$ implies

$$\|x(t)\| \leq C\|x_0\|e^{-\mu(t-t_0)}, t \geq t_0.$$

The solution $x = 0$ is asymptotically stable and the attraction is exponential in a
$\delta$-neighbourhood of $x = 0$.

**Proof**
From theorem 6.1 we have an estimate for the fundamental matrix of the equation

$$\dot{\Phi} = A\Phi, \Phi(t_0) = I.$$

As the eigenvalues of $A$ have all nonzero real part, there exist positive constants
$C$ and $\mu_0$ such that
$$\|\Phi(t)\| \leq Ce^{-\mu_0(t-t_0)}, t \geq t_0.$$
It follows from the assumptions on $f$ and $B$ that, for $\delta > 0$ sufficiently small, there
exists a constant $b(\delta)$ such that if $\|x\| < \delta$ we have

$$\|f(t, x)\| \leq b(\delta)\|x\| \, , \, t \geq t_0$$

and if $t_0$ is sufficiently large

$$\|B(t)\| \leq b(\delta) \, , \, t \geq t_0.$$

The existence and uniqueness theorem 1.1 yields that, in a neighbourhood of $x = 0$,
the solution of the initial value problem 7.1 exists for $t_0 \leq t \leq t_1$. It turns out
that this solution can be continued for all $t \geq t_0$.
The initial value problem 7.1 is equivalent with the integral equation (cf. the proof
of theorem 6.2)

(7.2) $$x(t) = \Phi(t)x_0 + \int_{t_0}^{t} \Phi(t - s + t_0)[B(s)x(s) + f(s, x(s))]ds.$$

Using the estimates for $\Phi$, $B$ and $f$ we have for $t_0 \leq t \leq t_1$

$$\|x(t)\| \leq \|\Phi(t)\|\|x_0\| + \int_{t_0}^{t} \|\Phi(t - s + t_0)\|[\|B(s)\|\|x(s)\|+$$

$$+\|f(s, x(s))\|]ds$$

$$\leq Ce^{-\mu_0(t-t_0)}\|x_0\| + \int_{t_0}^{t} Ce^{-\mu_0(t-s)}2b\|x(s)\|ds$$

so that

$$e^{\mu_0(t-t_0)}\|x(t)\| \leq C\|x_0\| + \int_{t_0}^{t} Ce^{-\mu_0(s-t_0)}2b\|x(s)\|ds.$$

We shall use now Gronwall's inequality in the form of theorem 1.2 ($\psi(s) = 2Cb$) to obtain

$$e^{-\mu_0(t-t_0)}\|x(t)\| \le C\|x_0\|e^{2Cb(t-t_0)}$$

or

(7.3) $$\|x(t)\| \le C\|x_0\|e^{(2Cb-\mu_0)(t-t_0)}.$$

If $\delta$ and consequently $b$ are small enough, the quantity $\mu = \mu_0 - 2Cb$ is positive and we have the required estimate for $t_0 \le t \le t_1$.
Now we shall choose $\|x_0\|$ such that

$$C\|x_0\| \le \delta.$$

So the estimate 7.3 holds for $t \ge t_0$. □

**Example 7.1** (oscillator with damping)

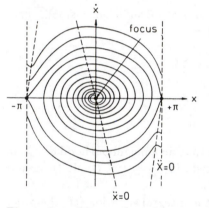

*Figure 7.1. Phase-plane for equation 7.4, $0 < \mu < 2$*

An important application is that, if one has a harmonic oscillator with linear damping, the stability of the equilibrium point does not change on adding nonlinear terms. Of course the domain of attraction changes. Consider for instance the mathematical pendulum with linear damping

(7.4) $$\ddot{x} + \mu\dot{x} + \sin x = 0 , \mu > 0.$$

We may consider the equation for the harmonic oscillator with damping in example 5.1 as the linearisation of equation 7.4 in a neighbourhood of (0,0). We can write the equation in vector form by transforming $x = x_1$, $\dot{x} = x_2$; we have

(7.5)
$$\begin{aligned} \dot{x}_1 &= x_2 \\ \dot{x}_2 &= -x_1 - \mu x_2 + (x_1 - \sin x_1) \end{aligned}$$

We find for the eigenvalues of the linearised equation

$$\lambda_{1,2} = -\frac{1}{2}\mu \pm \frac{1}{2}\sqrt{\mu^2 - 4}$$

so $Re\lambda_{1,2} < 0$. System 7.5 satisfies the requirements of the Poincaré-Lyapunov theorem so the equilibrium solution (0,0) is asymptotically stable. Of course we reach the same conclusion on generalising equation 7.4 to

(7.6) $$\ddot{x} + \mu\dot{x} + f(x) = 0, \ \mu > 0$$

with $f(x) = x +$ higher order terms in a neighbourhood of $x = 0$.

In the case that the linear part of the equation has periodic coefficients, we can apply the theory of Flocquet from section 6.3.

**Theorem 7.2**
Consider the equation in $\mathbb{R}^n$

(7.7) $$\dot{x} = A(t)x + f(t, x)$$

with $A(t)$ a $T$-periodic, continuous matrix; the vector function $f(t, x)$ is continuous in $t$ and $x$ and Lipschitz-continuous in $x$ for $t \in \mathbb{R}$, $x$ in a neighbourhood of $x = 0$. Moreover we have

$$\lim_{\|x\| \to 0} \frac{\|f(t, x)\|}{\|x\|} = 0 \text{ uniformly in } t.$$

If the real parts of the characteristic exponents of the linear periodic equation

(7.8) $$\dot{y} = A(t)y$$

are negative, the solution $x = 0$ of equation 7.7 is asymptotically stable. Also the attraction is exponential in a $\delta$-neighbourhood of $x = 0$.

**Proof**
We shall use remark 1 from section 6.3 by transforming

$$x = P(t)z$$

with $P(t)$ a periodic matrix belonging to the fundamental matrix solution of equation 7.8. We find
(7.9) $$\dot{z} = Bz + P^{-1}(t)f(t, P(t)z).$$

The constant matrix $B$ has only eigenvalues with negative real parts. The solution $z = 0$ of equation 7.9 satisfies the requirements of the Poincaré-Lyapunov theorem from which follws the result. $\qquad \square$

**Remark 1**

Anticipating to the Poincaré-Lyapunov theorem we have formulated theorem 3.1 on the relation between linear and nonlinear equations in the case of attraction. Theorem 3.1 is a special case of theorem 7.1.

**Remark 2**

In example 5.3 we have shown that positive attraction by a critical point does not always imply stability. Theorem 7.1 tells us that if, within the conditions of the theorem, we find positive attraction in the linear approximation, the solution of the nonlinear problem is asymptotically stable.

The formulation of theorem 7.1 contains conditions which are sharp with respect to the result required. One could think for instance that if $y = 0$ is a positive attractor of the linear equation

$$\dot{y} = Ay + B(t)y$$

that on adding a smooth nonlinear term, the trivial solution of the nonlinear equation is asymptotically stable. The following ingenious counter example shows that the condition

$$\lim_{t \to 0} \|B(t)\| = 0$$

is essential.

**Example 7.2 (Perron)**

Consider for $t \geq 1$ the system

$$
\begin{aligned}
\dot{x}_1 &= -ax_1 \\
\dot{x}_2 &= [-2a + \sin(lnt) + \cos(lnt)]x_2 + x_1^2
\end{aligned}
$$

with constant $a > \frac{1}{2}$. The conditions of the theorem of Poincaré and Lyapunov have not been satisfied. On linearising in a neighbourhood of the trivial solution we find

$$
\begin{aligned}
\dot{y}_1 &= -ay_1 \\
\dot{y}_2 &= [-2a + \sin(lnt) + \cos(lnt)]y_2
\end{aligned}
$$

with independent solutions

$$
\begin{aligned}
y_1(t) &= e^{-at} \\
y_2(t) &= e^{t\sin(lnt)-2at}.
\end{aligned}
$$

These solutions tend to zero as $t \to \infty$. Substitution of the solution $x_1(t)$ into the second equation yields a linear inhomogeneous equation.

Using the method of variation of constants we can compute the solution $x_2(t)$. We find, with $c_1$ and $c_2$ constants

$$\begin{aligned} x_1(t) &= c_1 e^{-at} \\ x_2(t) &= e^t \sin(\ln t) - 2at\left(c_2 + c_1^2 \int_0^t e^{-\tau} \sin(\ln \tau) d\tau\right). \end{aligned}$$

Consider the sequence $t_n = \exp\left((2n + \frac{1}{2})\pi\right)$, $n = 1, 2, \cdots$. We estimate

$$\int_0^t e^{-\tau} \sin(\ln \tau) d\tau > \int_{t_n e^{-\pi}}^{t_n e^{-2\pi/3}} e^{-\tau} \sin(\ln \tau) d\tau >$$

$$> t_n (e^{-2\pi/3} - e^{-\pi}) e^{\frac{1}{2} e^{-\pi} t_n}.$$

Choosing $a$ such that

$$1 < 2a < 1 + \frac{1}{2} e^{-\pi}$$

we find that the solutions are not bounded as $t \to \infty$ unless $c_1 = 0$.

## 7.2  Instability of the trivial solution

When analysing a system linearised in a neighbourhood of an equilibrium solution, one can also conclude under certain circumstances that the solution is unstable. Theorem 3.2 has been formulated to deal with this case and we shall now formulate and prove a more general version of this result.

**Theorem 7.3**
Consider the equation in $\mathbb{R}^n$

(7.10) $$\dot{x} = Ax + B(t)x + f(t, x) \qquad , t \geq t_0.$$

$A$ is a constant $n \times n$-matrix with eigenvalues of which at least one has positive real part; $B(t)$ is a continuous $n \times n$-matrix with the property

$$\lim_{t \to \infty} \|B(t)\| = 0.$$

The vector function $f(t, x)$ is continuous in $t$ and $x$, Lipschitz- continuous in $x$ in a neighbourhood of $x = 0$; if moreover we have

$$\lim_{\|x\| \to 0} \frac{\|f(t, x)\|}{\|x\|} = 0, \qquad \text{uniformly in } t.$$

the trivial solution of equation 7.10 is unstable.

**Proof**

First we shall transform equation 7.10 using a non-singular constant $n \times n$-matrix $S : x = Sy$. Equation 7.10 becomes

(7.11) $$\dot{y} = S^{-1}ASy + S^{-1}B(t)Sy + S^{-1}f(t, Sy).$$

The solution $x(t)$ is real-valued, $y(t)$ will generally be a complex function. Instability of the trivial solution of equation 7.11 implies instability of the trivial solution of equation 7.10. For simplicity we assume that $S$ can be chosen such that $S^{-1}AS$ is in diagonal form, i.e. the eigenvalues $\lambda_i$ of the matrix $A$ can be found on the main diagonal of $S^{-1}AS$ and the other matrix elements are zero. For the more general case see Coddington and Levinson (1955), chapter 13.1. We put

$$\begin{aligned} Re(\lambda_i) &\geq \sigma > 0 & ,i = 1, \cdots, k \\ Re(\lambda_i) &\leq 0 & ,i = k+1, \cdots, n. \end{aligned}$$

We introduce furthermore the quantities

$$R^2 = \sum_{i=1}^{k} |y_i|^2 \text{ and } r^2 = \sum_{i=k+1}^{n} |y_i|^2.$$

Using equation 7.11 we compute the derivatives of $R^2$ and $r^2$; we shall use

$$\begin{aligned} \frac{d}{dt}|y_i|^2 &= \frac{d}{dt}(y_i\bar{y}_i) = \dot{y}_i\bar{y}_i + y_i\dot{\bar{y}}_i \\ &= 2Re\lambda_i|y_i|^2 + (S^{-1}B(t)Sy)_i\bar{y}_i + y_i(S^{-1}B(t)Sy)_i + \\ & \quad +(S^{-1}f(t, Sy))_i\bar{y}_i + y_i(S^{-1}f(t, Sy))_i. \end{aligned}$$

Now we may choose $\varepsilon > 0$, $\delta_0$ and $\delta$ such that for $t \geq t_0$ and $\|y\| \leq \delta$ we have

$$|S^{-1}B(t)Sy|_i \leq \varepsilon|y_i| \ , \ |S^{-1}f(t, Sy)_i| \leq \varepsilon|y_i|.$$

So we find

$$\frac{1}{2}\frac{d}{dt}(R^2 - r^2) \geq \sum_{i=1}^{k}(Re\lambda_i - \varepsilon)|y_i|^2 - \sum_{i=k+1}^{n}(Re\lambda_i + \varepsilon)|y_i|^2.$$

If we choose $0 < \varepsilon \leq \frac{1}{2}\sigma$ we have

$$\begin{aligned} (Re\lambda_i - \varepsilon) &\geq \sigma - \varepsilon \geq \varepsilon & i = 1, \cdots, k \\ (Re\lambda_i + \varepsilon) &\leq \varepsilon & i = k+1, \cdots, n. \end{aligned}$$

It follows that

(7.12) $$\frac{1}{2}\frac{d}{dt}(R^2 - r^2) \geq \varepsilon(R^2 - r^2) \ , \ t \geq t_0, \ \|y\| \leq \delta.$$

If we choose the initial values such that

$$(R^2 - r^2)_{t=t_0} = a > 0$$

we find with 7.12

$$\|y\|^2 \geq R^2 - r^2 \geq ae^{2\epsilon(t-t_0)}.$$

So this solution leaves the domain determined by $\|y\| \leq \delta$; the trivial solution is unstable. $\qquad\qquad\qquad\qquad\qquad\qquad\qquad\qquad\qquad\qquad\qquad\qquad$ □

We can relate the result of theorem 7.3 with theorem 3.3 in the autonomous case $B(t) = 0$, $\delta f/\delta t = 0$, $t \geq t_0$. In this case theorem 3.3 guarantees the existence of unstable manifolds corresponding with the eigenvalues with positive real part. The solutions which are leaving a neighbourhood of the trivial solution can be found in these unstable manifolds. In the more general non-autonomous case of equation 7.10 the solutions which are leaving a neighbourhood of the trivial solution have not such a convenient and nice geometric characterisation.

The results of theorems 7.1 and 7.3 will be illustrated with a population model of mathematical biology.

**Example 7.3** (competing species)
Suppose that two animal species are living in a certain territory; the population densities are $x$ and $y$, $x, y \geq 0$. The growth rate of the two species (difference between birth rate and death rate) is 1. If the population densities increase there are natural limits to their growth and if there is no interaction (competition) between the species the equations are

$$\dot{x} = x - ax^2 \,,\; \dot{y} = y - by^2 \,,\; a, b > 0.$$

These are the logistic equations or Verhulst equations, called after the $19^{th}$ century Belgian mathematician P.F. Verhulst.
We shall now add the following assumption: although the species with density $y$ is not actually predating on the other species $(x)$, encounters between the two species are always to the advantage of $y$. One can think for instance of the coexistence in a territory of birds of prey which are of different size like the tawny owl $(y)$ and the long-eared owl $(x)$ or the long-eared owl $(y)$ and the little owl $(x)$.
The equations become with this assumption added

(7.13)
$$\begin{aligned} \dot{x} &= x - ax^2 - cxy \quad, x \geq 0 \\ \dot{y} &= y - by^2 + dxy \quad, y \geq 0, a, b, c, d > 0. \end{aligned}$$

There are the following critical points corresponding with equilibrium solutions:

$$(0,0),\; (0,1/b),\; (1/a,0),\; \left( \frac{b-c}{ab+cd}, \frac{a+d}{ab+cd} \right).$$

If $b \geq c$, the fourth critical point is also found in the domain which interests us. In characterising critical points on the boundary, we treat them formally as interior points. Analysis by linearisation, cf. sections 2.3 and 3.1, of the critical points produces the following results: If $b > c$

| | | | |
|---|---|---|---|
| $(0,0)$ | $\lambda_1 = 1, \lambda_2 = 1$ | node | unstable |
| $(0, 1/b)$ | $\lambda_1 = 1 - c/b, \lambda_2 = -1$ | saddle | unstable |
| $(1/a, 0)$ | $\lambda_1 = -1, \lambda_2 = 1 + d/a$ | saddle | unstable |
| $\left(\frac{b-c}{ab+cd}, \frac{a+d}{ab+cd}\right)$ | $Re\lambda_1, \lambda_2 < 0$ | node or focus | stable |

If $b < c$

| | | | |
|---|---|---|---|
| $(0,0)$ | $\lambda_1 = 1, \lambda_2 = 1$ | node | unstable |
| $(0, 1/b)$ | $\lambda_1 = 1 - c/b, \lambda_2 = -1$ | node | stable |
| $(1/a, 0)$ | $\lambda_1 = -1, \lambda_2 = 1 + d/a$ | saddle | unstable |

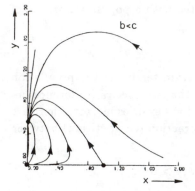

Figure 7.2  *strong interaction*

In this list we have omitted some transition cases like $b = c$. From the theorems 7.1 and 7.3 we conclude that in this list the stability by linear analysis carries over to the full nonlinear system. We also conclude that this model can describe two rather different situations. If $b < c$ the species $x$ dies out, we shall call this strong interaction between the species. If $b > c$ then there exists a positive equilibrium solution with coexistence of the two species; we shall call this the case of weak interaction. In this second case there are, on approaching the equilibrium solution, qualitatively two possibilities: node or focus.

Note that this behaviour of the solutions for different values of the parameters is apriori or intuitively not obvious. We can add to this discussion a quantitative characterisation of the stable and unstable manifolds of the saddle points which arise, see also theorem 3.3. This is left to the reader.

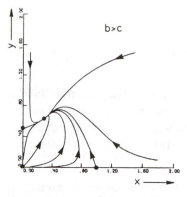

*Figure 7.3 weak interaction*

## 7.3 Stability of periodic solutions of autonomous equations

We have shown in section 5.4 how, by a shifting procedure, we can reduce the problem of stability of a periodic solution to the problem of stability of the trivial solution. The equation which arises after this shifting prodecure, will in many cases be of the form 7.7, so that, in the case of suspected stability, we shall try to apply theorem 7.2. If we want to demonstrate instability, the idea of transforming in the proof of theorem 7.2 in combination with theorem 7.3 would be the advised thing to do.

A new problem arises when studying a periodic solution of the autonomous equation

$$(7.14) \qquad \dot{x} = f(x).$$

We have shown in section 5.4, that if $\phi(t)$ is a $T$-periodic solution of the autonomous equation 7.13, linearisation in a neighbourhood of the periodic solution produces a linear periodic system with as one of its solutions $\dot{\phi}(t)$. This implies that at least one of the real parts of the characteristic exponents is zero, so that theorem 7.2 does not apply.

This situation happens quite often and to deal with it we can formulate the following result.

**Theorem 7.4**

Consider equation 7.13 which has a $T$-periodic solution $\phi(t)$; $f(x)$ is continuously differentiable in a domain in $\mathbb{R}^n$, $n > 1$, containing $\phi(t)$. Suppose that linearisation of equation 7.13 in a neighbourhood of $\phi(t)$ yields the equation

$$\dot{y} = \frac{\delta f}{\delta x}(\phi(t))y$$

with characteristic exponents of which one has real part zero and $n - 1$ exponents have real parts negative. Then there exists a constant $\varepsilon > 0$ such that on starting a solution $x(t)$ of equation 7.13 in a neighbourhood of $\phi(t)$ with

$$\|x(t_1) - \phi(t_0)\| < \varepsilon \text{ for certain } t_0 \text{ and } t_1$$

we have

$$\lim_{t\to\infty} \|x(t) - \phi(t + \theta_0)\| = 0$$

in which $\theta_0$ is a certain constant, the asymptotic phase.

**Proof**
It is possible to demonstrate the existence of a $(n - 1)$-dimensional manifold in a neighbourhood of $\phi(t)$ which has the property that on starting the solutions in this manifold, these solutions all tend to $\phi(t)$. The proof is not difficult but rather lenghty; we refer the reader to Coddington and Levinson (1955), chapter 13.2, theorem 2.2. □

In some books, a periodic solution which is stable in the sense of theorem 7.4, has been called *orbitally stable*.
The theorems which we discussed in this chapter are the basic results of the theory of stability by linearisation. The literature in this field is very extensive and we refer to Coddington and Levinson (1955), Hale (1969), Sansone and Conti (1964), Cesari (1971), Roseau (1966) and Hahn (1967).
We conclude with some examples.

**Example 7.4** (generalised Liénard equation)
In an application of theorem 6.6 we considered the equation

$$\ddot{x} + f(x)\dot{x} + g(x) = 0.$$

Suppose this equation has a $T$-periodic solution $x = \phi(t)$. Using the calculation in this application together with theorem 7.4 we conclude that the periodic solution is stable if

$$\int_0^T f(\phi(t))dt > 0.$$

**Example 7.5**
Consider the two-dimensional system

(7.15)
$$\dot{x} = f(x,y)$$
$$\dot{y} = g(x,y)$$

with $T$-periodic solution $x = \phi(t)$, $y = \psi(t)$. We use the shifting procedure by transforming $x = \phi(t) + u$, $y = \psi(t) + v$ to obtain after expansion

$$\dot{u} = f_x(\phi(t), \psi(t))u + f_y(\phi(t), \psi(t))v + \cdots$$
$$\dot{v} = g_x(\phi(t), \psi(t))u + g_y(\phi(t), \psi(t))v + \cdots$$

where $\cdots$ is short for the nonlinear terms in $u$ and $v$. Using theorems 6.6 and 7.4 we conclude that this periodic solution of system 7.4 is stable if

$$\int_0^T [f_x(\phi(t), \psi(t)) + g_y(\phi(t), \psi(t))]dt < 0.$$

One can check that this result applies to the problem of example 7.4.

## 7.4  Exercises

7-1. We extend the model for competing species in example 7.3 by adding a third species. Think for instance of the coexistence in a territory of the tawny owl ($z$), the long-eared owl ($y$) and the little owl ($x$).
The equations are for $x, y, z \geq 0$

$$
\begin{aligned}
\dot{x} &= x - a_1 x^2 - a_2 xy - a_3 xz \\
\dot{y} &= y - b_1 y^2 + b_2 xy - b_3 yz \\
\dot{x} &= z - c_1 z^2 + c_2 xz + c_3 yz
\end{aligned}
$$

The coefficients are all positive. Study the equilibrium solutions of this system by linear analysis and also with respect to their nonlinear stability.

7-2. The equation
$$
\ddot{x} - (1 - x^2 - \dot{x}^2)\dot{x} + x = 0
$$
has a critical point and a limit cycle solution.

a. Determine the critical point and characterise it; determine the limit cycle.

b. Establish the stability of the periodic solution.

7-3. Consider the system
$$
\begin{aligned}
\dot{x} &= 1 + y - x^2 - y^2 \\
\dot{y} &= 1 - x - x^2 - y^2.
\end{aligned}
$$

a. Determine the critical points and characterise them.

b. Show that if a periodic solution exists, its corresponding cycle intersects the line $y = -x$ in the phase-plane.

c. Show that the system has a periodic solution (use polar coordinates).

d. Linearise the system near this periodic solution and determine the characteristic exponents.

e. Determine the stability of the periodic solution found in c.

7-4. The model for competing species in example 7.3 can be changed by assuming that the interaction is to the disadvantage of both species. We have

$$
\begin{aligned}
\dot{x} &= x - x^2 - axy \quad , x \geq 0 \\
\dot{y} &= y - y^2 - axy \quad , y \geq 0
\end{aligned}
$$

with $a$ a positive constant.

a. Determine the critical points and characterise them by linear analysis.

b. What are the attraction properties of the critical points in the full nonlinear system?

c. Show that a solution starting in the domain $x \geq 0, y \geq 0$ remains in this domain.

d. Sketch the phase-flow for the cases $0 < a < 1, a = 1, a > 1$.

e. Determine the conditions for extinction (as $t \to \infty$) of one of the species, extinction of both species, coexistence of both species; assume in all these cases $x(0) > 0, y(0) > 0$.

7-5. Consider the three-dimensional system

$$\dot{x} = (1-z)[(4-z^2)(x^2+y^2-2x+y) - 4(-2x+y) - 4]$$
$$\dot{y} = (1-z)[(4-z^2)(xy-x-zy) - 4(-x-zy) - 2z]$$
$$\dot{z} = z^2(4-z^2)(x^2+y^2)$$

Determine the equilibrium solutions and their stability properties.

7-6. We are interested in the stability properties of the solution $(0,0)$ of the system

$$\dot{x} = \alpha x + y^n$$
$$\dot{y} = \alpha y - x^n$$

with $\alpha \in \mathbb{R}, n \in \mathbb{N}$.

a. What are these properties if $\alpha \neq 0$?

b. What happens if $\alpha = 0$?

7-7. An extension of the Volterra-Lotka equations involves one prey (population density $x$) and two predators $(y, z)$.

$$\dot{x} = ax - xy - xz$$
$$\dot{y} = -by + xy$$
$$\dot{x} = -cz + xz \qquad , x \geq 0, y \geq 0, z \geq 0, a, b, c \text{ positive parameters.}$$

a. Characterise the equilibrium solutions by linear analysis.

b. Do periodic solutions exist with $y(t)z(t) > 0$?

c. Is it possible that, starting with $y(0)z(0) > 0$, one of the predators becomes extinct?

# 8 Stability analysis by the direct method

## 8.1 Introduction

In this chapter we shall discuss a method for studying the stability of a solution, which is very different from the method of linearisation of the preceding chapter. When linearising one starts off with small perturbations of the equilibrium or periodic solution and one studies the effect of these *local* perturbations. In the so-called direct method one characterises the solution in a way with respect to stability which is not necessarily local.

The method originates from the field of classical mechanics where this non-local characterisation arises from the laws of statics and dynamics. A basic idea can be found in the work of Torricelli (1608-1647), a student of Galileï. From Torricelli we have the following "axiom", as he calls it: "Connected heavy bodies cannot start moving by themselves if their common centre of gravity does not move downward." Torricelli's axiom finds its application in statics; Huygens (1629-1695) developed this idea, together with other insights, for the dynamics of particles, bodies and fluids. Later this was taken up by Lagrange (1736-1813) who formulated his well-known principle for the stability of a mechanical system:

"A mechanical system which is in a state where its potential energy has an isolated minimum, is in a state of stable equilibrium."

We shall prove the validity of the principle of Lagrange in section 8.3; we shall also illustrate the concept of potential energy there. The reader who is interested in the historical development of mechanics should consult the classic by Dijksterhuis (1950).

Around 1900 these ideas of stability in mechanics were generalised strongly by Lyapunov. In his work differential equations are studied which have not been characterised apriori by a potential energy or a quantity energy in general. To introduce Lyapunov's ideas we discuss an example in which also the geometric aspects of the method are transparent.

Consider the system of equations

(8.1)
$$\begin{aligned} \dot{x} &= ax - y + kx(x^2 + y^2) \\ \dot{y} &= x - ay + dy(x^2 + y^2) \end{aligned}$$

with constants $a$ and $k$, $a > 0$. We are interested in the stability of the trivial solution.

Linearisation in a neighbourhood of $(0,0)$ yields the eigenvalues

$$\pm(a^2 - 1)^{\frac{1}{2}},$$

so in the linear approximation we find for the critical point $(0,0)$

$$
\begin{array}{ll}
a^2 > 1 & \text{saddle} \\
a^2 = 1 & \text{degenerate case} \\
0 < a^2 < 1 & \text{centre.}
\end{array}
$$

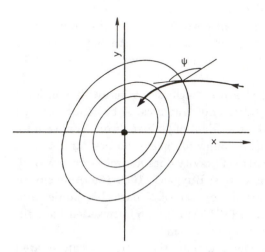

*Figure 8.1*

We conclude with theorem 7.3 that if $a^2 > 1$, the trivial solution is unstable. If $0 < a^2 \leq 1$ the method of linearisation of chapter 7 is not conclusive.
We now consider a one-parameter family of ellipses around $(0,0)$:

$$x^2 - 2axy + y^2 = c.$$

The parameter is $c$, $a$ is fixed with $0 < a^2 < 1$.
An orbit in the $x,y$-phaseplane, corresponding with a solution of system 8.1, will intersect an ellips with angle $\psi$ (this angle is taken between the tangent vector of the orbit and the outward directed normal vector in the point of intersection). If $\frac{\pi}{2} < \psi < 3\frac{\pi}{2}$, $\cos \psi < 0$, the orbit enters the interior of this particular ellipse. If $\cos \psi < 0$ for all solutions and all ellipses in a neighbourhood of $(0,0)$, the trivial solution is asymptotically stable. We compute $\cos \psi$.
The tangent vector $\vec{\tau}$ of the orbit is given by $\vec{\tau} = (\dot{x}, \dot{y})$. The gradient vector of the function

$$V = x^2 - 2axy + y^2$$

produces the normal vector field

$$\nabla V = \left( \frac{\partial V}{\partial x}, \frac{\partial V}{\partial y} \right) = (2x - 2ay, -2ax + 2y).$$

We compute $\cos \psi$ using the vector product

$$\cos \psi = \frac{(\nabla V . \vec{\tau})}{\|\nabla V\| . \|\vec{\tau}\|}$$

$$= \frac{\frac{\partial V}{\partial x}\dot{x} + \frac{\partial V}{\partial y}\dot{y}}{\|\nabla V\| . \|\vec{\tau}\|}$$

The sign of $\cos \psi$ is determined by the numerator

$$\frac{\partial V}{\partial x}\dot{x} + \frac{\partial V}{\partial y}\dot{y} = L_t V.$$

In section 2.4 we have called $L_t$ the orbital derivative of the function $V$. Using equations 8.1 we find in this case

$$\begin{aligned} L_t V &= (2x - 2ay)\dot{x} + (-2ax + 2y)\dot{y} \\ &= 2k(x^2 + y^2)(x^2 + y^2 - 2axy). \end{aligned}$$

$$\begin{aligned} \text{So} \quad \cos \psi &< 0 \quad if \quad k < 0 \\ \cos \psi &> 0 \quad if \quad k > 0. \end{aligned}$$

The result holds for all ellipses and all orbits in a neighbourhood of $(0,0)$ so we conclude that the trivial solution is asymptotically stable if $k < 0$, unstable if $k > 0$. The result holds for all orbits so that for $k < 0$ and $0 < a^2 < 1$ we have even global stability of $(0,0)$.
The case $a^2 = 1$ is left to the reader.

In this chapter we shall present the most important results of the direct method of Lyapunov; many more results can be found in Hahn (1967). The direct method has also been applied and extended considerably in the theory of optimal control.

## 8.2  Lyapunov functions

Consider the equation

(8.2) $$\dot{x} = f(t, x), \quad t \geq t_0, x \in D \subset \mathbf{R}^n.$$

and assume that the trivial solution satisfies the equation, so $f(t, 0) = 0$, $t \geq t_0$, $0 \in D$.

**Definition** $V(t, x)$

In this chapter the function $V(t, x)$ is defined and continuously differentiable in $[t_0, \infty) \times D$, $D \subset \mathbb{R}^n$. Moreover $x = 0$ is an interior point of $D$ and

$$V(t, 0) = 0.$$

In some cases the function $V(t, x)$ does not depend explicitly on $t$ and we write for short $V(x)$. The function $V(t, x)$ being positively definite or negatively definite is introduced as follows.

**Definition**

The function $V(x)$ (with $V(0) = 0$) is called positively (negatively) definite in $D$ if $V(x) > 0$ ($< 0$) for $x \in D$, $x \neq 0$.

There are cases in which the function $V(x)$ takes the value zero in a subset of $D$ but has otherwise a definite sign.

**Definition**

The function $V(x)$ (with $V(0) = 0$) is called positively (negatively) semidefinite in $D$ if $V(x) \geq 0$ ($\leq 0$) for $x \in D$.

If the function $V(t, x)$ depends explicitly on $t$, these definitions are adjusted as follows.

**Definition**

The function $V(t, x)$ is called positively (negatively) definite in $D$ if there exists a function $W(x)$ with the following properties: $W(x)$ is defined and continuous in $D$, $W(0) = 0$, $0 < W(x) \leq V(t, x)$ ($V(t, x) \leq W(x) < 0$) for $x \neq 0$, $t \geq t_0$. To define semidefinite functions $V(t, x)$ we replace $<$ ($>$) by $\leq$ ($\geq$).

**Example 8.1**

Definite functions which are used very often are quadratic functions with positive coefficients. Consider in $\mathbb{R}^3$ $D = \{(x, y,) \mid x^2 + y^2 + z^2 \leq 1\}$ and for $t \geq 0$ the functions

$$
\begin{array}{ll}
x^2 + 2y^2 + 3z^2 + z^3 & \text{positive definite} \\
x^2 + z^2 & \text{positive semidefinite} \\
-x^2 \sin^2 t - y^2 - 4z^2 & \text{negative semidefinite} \\
x^2 + y^2 + \cos^3 tz^2 & \text{not sign definite}
\end{array}
$$

In the sequel we shall use a simple extension of the concept of orbital derivative, cf. section 2.4.

**Definition**
The orbital derivative $L_t$ of the function $V(t, x)$ in the direction of the vectorfield $x$, where $x$ is a solution of equation 8.2 $\dot{x} = f(t, x)$ is

$$\begin{aligned} L_t V &= \frac{\partial V}{\partial t} + \frac{\partial V}{\partial x}\dot{x} &&= \frac{\partial V}{\partial t} + \frac{\partial V}{\partial x}f(t, x) \\ &&&= \frac{\partial V}{\partial t} + \frac{\partial V}{\partial x_1}f_1(t, x) + \cdots + \frac{\partial V}{\partial x_n}f_n(t, x) \end{aligned}$$

with $x = (x_1, \cdots, x_n)$ and $f = (f_1, \cdots, f_n)$.

Now we can formulate and prove the basic theorems.

**Theorem 8.1**
Consider equation 8.2 $\dot{x} = f(t, x)$ with $f(t, 0) = 0$, $x \in D \subset \mathbb{R}^n$, $t \geq t_0$. If a function $V(t, x)$ can be found, defined in a neighbourhood of $x = 0$ and positively definite for $t \geq t_0$ with orbital derivative negatively semidefinite, the solution $x = 0$ is stable in the sense of Lyapunov.

**Proof**
In a neighbourhood of $x = 0$ we have for certain $R > 0$ and $\|x\| \leq R$

$$V(t, x) \geq W(x) > 0, \ x \neq 0, \ t \geq t_0$$

$$L_t V \leq 0.$$

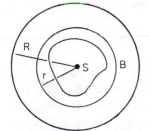

*Figure 8.2*

Consider a spherical shell $B$, given by $0 < r \leq \|x\| \leq R$ and put

$$m = \min_{x \in B} W(x).$$

Consider now a neighbourhood $S$ of $x = 0$ with the property that if $x \in S$, $V(t, x) < m$. Such a neighbourhood exists as $V(t, x)$ is continuous and positively definite while $V(t,0) = 0$. Starting a solution in $S$ at $t = t_0$, the solution can never enter $B$, as we have for $t \geq t_0$

$$V(t, x(t)) - V(t_0, x(t_0)) = \int_{t_0}^t L_\tau V(\tau, x(\tau))d\tau \leq 0.$$

In other words, the function $V(t, x(t))$ cannot increase along a solution and this would be necessary to enter $B$ as initially $V(t_0, x(t_0)) < m$.

We can repeat the argument for arbitrary small $R$, from which follows the stability. $\qquad\square$

The function $V(t, x)$ which we employed in this theorem is called a Lyapunov-function. Whether such functions exist and how one should construct them is known in a number of cases but not in general. For each class of problems we have to start again if we want to use the concept of Lyapunov-function.

Note that in the formulation of theorem 8.1 we have assumed that the orbital derivative is semidefinite. This includes the case that $L_t V = 0$, $t \geq t_0$, $x \in D$, in other words: $V(t, x)$ is a first integral of the equation; see also section 2.4. We shall return to the part played by first integrals in the applications of section 8.3. In requiring more of the orbital derivative, we obtain a stronger form of stability.

**Theorem 8.2**
Consider equation 8.2 $\dot{x} = f(t, x)$ with $f(t, 0) = 0$, $x \in D \subset \mathbf{R}^n$, $t \geq t_0$. If a function $V(t, x)$ can be found, defined in a neighbourhood of $x = 0$, which for $t \geq t_0$ is positively definite in this neighbourhood with negative definite orbital derivative, the solution $x = 0$ is asymptotically stable.

**Proof**
It follows from theorem 8.1 that $x = 0$ is a stable solution. Is it possible that for each $R > 0$ there exists a solution $x(t)$ which starts in the domain given by $\|x\| \leq R$ and which does not tend to zero? Put differently: is there a solution $x(t)$ and a constant $a > 0$ such that $\|x(t)\| \geq a$ for $t \geq t_0$, when starting arbitrarily close to zero?

Suppose this is the case, the solution remains in the spherical shell $B$: $a \leq \|x(t)\| \leq R$, $t \geq t_0$. We have $L_t V(t, x) \leq W(x) < 0$, $x \neq 0$. So we have in $B$

$$L_t V \leq -\mu, \ \mu > 0$$

so that
$$V(t, x(t)) - V(t_0, x(t_0)) = \int_{t_0}^t L_\tau V(\tau, x(\tau))d\tau \leq -\mu(t - t_0).$$

On the other hand, we know that $V(t, x)$ is positively sign definite, whereas it follows from this estimate that after some time $V(t, x)$ becomes negative. This is a contradiction. $\qquad\square$

In the proofs of theorems 8.1 and 8.2 we have developed, while using the Lyapunov-function $V(t,x)$, a picture of the behaviour of the solutions. In particular in the proof of theorem 8.2 we have indicated when a solution has left the spherical shell $B$ and one can compute how long this will take at most.

If the trivial solution is asymptotically stable, we are interested of course in the set of all initial values corresponding with solutions which go to zero. This set is called the domain of attraction; we shall define this set for autonomous equations.

## Definition

Consider the equation $\dot{x} = f(x)$ and suppose that $x = 0$ is an asymptotically stable solution. A set of points $x_0$ with the property that for the solution of

$$\dot{x} = f(x) \ , \ x(0) = x_0$$

we have $x(t) \to 0$ for $t \to \infty$, is called a domain of attraction of $x = 0$.

Using a Lyapunov-function, we are now able to characterise domains of attraction. The following result follows directly from the proof of theorem 8.2.

## Corollary theorem 8.2.

Consider equation $\dot{x} = f(x)$ in $\mathbb{R}^n$ with $f(0) = 0$. The Lyapunov- function $V(x)$ is positively definite for $\|x\| \leq R$. $S$ is a closed $(n-1)$-dimensional manifold which encloses $x = 0$ and which is contained in the ball with radius $R$. Suppose that

$$\begin{aligned} L_t V &< 0 \ , \ x \text{ in the interior of } S \\ L_t V &= 0 \ , \ x \in S \\ L_t V &> 0 \ , \ x \text{ outside } S. \end{aligned}$$

Then each manifold which encloses $x = 0$ and which is contained in the interior of $S$, bounds a domain of attraction of $x = 0$.

A more extensive treatment of results for domains of attraction can be found in Hahn (1967), sections 4.26 and 4.33.

Using a Lyapunov-function one can also establish the instability of a solution.

## Theorem 8.3

Consider equation 8.2 $\dot{x} = f(t,x)$ with $f(t,0) = 0$, $x \in D \subset \mathbb{R}^n$, $t \geq t_0$. If there exists a function $V(t,x)$ in a neighbourhood of $x = 0$ such that:

a. $V(t,x) \to 0$ for $\|x\| \to 0$, uniformly in $t$;

b. $L_t V$ is positively definite in a neighbourhood of $x = 0$;

c. from a certain value $t = t_1 \geq t_0$, $V(t, x)$ takes positive values in each sufficiently small neighbourhood of $x = 0$;

then the trivial solution is unstable.

**Proof**

For certain positive constants $a$ and $b$ we have with $x \neq 0$ and $\|x\| \leq a$: $L_t V(t, x) \geq W(x) > 0$ and $|V(t, x)| \leq b$; the last estimate follows from assumption a.
Suppose that $x = 0$ is a stable solution. Then there exists a $\varepsilon > 0$ with $0 < \varepsilon < a$ such that when starting in $x_0$ with $\|x_0\| \leq \varepsilon$, we have $\|x(t)\| \leq a$ for $t \geq t_1$. Using assumption c we can choose $x_0$ such that $V(t_1, x_0) > 0$. We find for the solution $x(t)$ which starts in $x_0$ at $t = t_1$:

$$V(t, x(t)) - V(t_1, x_0) = \int_{t_1}^{t} L_\tau V(\tau, x(\tau)) > 0.$$

So $V(t, x(t))$ is non-decreasing. Consider now the set of points $x$ with the property that $V(t, x) \geq V(t_1, x_0)$ and $\|x\| \leq a$. This set is contained in the spherical shell $S$ given by $0 < r \leq \|x\| \leq a$. We have

$$\mu = \inf_S W(x) > 0$$

so that

$$V(t, x(t)) - V(t_1, x_0) \geq \mu(t - t_1).$$

So for $\|x\| \leq a$, $V(t, x)$ can become arbitrarily large; this is a contradiction. □

Theorem 8.3 has been generalized by Chetaev in such a way that its applicability has increased; see again Hahn (1967).
In the next sections we study some applications and examples.

## 8.3   Hamiltonian systems and systems with first integrals

In example 2.12 we introduced Hamilton's equations:

$$(8.3) \qquad p_i = -\frac{\partial H}{\partial q_i} \, , \; \dot{q}_i = \frac{\partial H}{\partial p_i} \, , \; i = 1, \cdots, n,$$

in which $H$ is a twice continuously differentiable function of the $2n$ variables $p_i$ and $q_i$, $H : \mathbb{R}^{2n} \to \mathbb{R}$. The Hamilton function $H$ is a first integral of the equations, i.e. the orbital derivative produces

$$L_t H = 0.$$

Suppose that the trivial solution satisfies system 8.3 and that we are interested in the stability of this equilibrium solution. Without loss of generality we may assume $H(0,0) = 0$ for we can add a constant to $H(p,q)$ without changing system 8.3. If now we require $H(p,q)$ to be sign definite in a neighbourhood of $x = 0$, the stability of the trivial solution follows by application of Lyapunov's theorem 8.1. The Hamilton function $H$ is also a Lyapunov function. So we have the following result.

## Theorem 8.4

Consider Hamilton's equations 8.3, of which we assume that they admit the trivial solution. If $H(p,q) - H(0,0)$ is sign definite in a neighbourhood of $(p,q) = (0,0)$, the trivial solution is stable in the sense of Lyapunov.

## Remark

In section 2.4 we discussed special but frequently occurring Hamiltonian systems, in which $(0,0)$ is a nondegenerate (Morse) critical point of the Hamilton function. We made use of the Morse lemma (see appendix 1); this application leads off to the existence of invariant, closed manifolds around the critical point in phase space. From this geometric picture one can also deduce stability.

Mechanical systems in which the force field can be derived from a potential $\phi(q)$ are characterised in many cases by the Hamiltonian

$$(8.4) \qquad H(p,q) = \frac{1}{2} \sum_{i=1}^{n} p_1^2 + \phi(q).$$

The equations of motion 8.3 can be written as

$$(8.5) \qquad \ddot{q}_i = -\frac{\partial \phi}{\partial q_i} \ , \ i = 1, \cdots, n.$$

Equilibrium solutions correspond with critical points determined by

$$p_i(= \dot{q}_i) = 0 \ , \ \frac{\partial \phi}{\partial q_i} = 0 \ , \ i = 1, \cdots, n.$$

It follows from theorem 8.4 that the equilibrium solution is stable if $\phi(q)$ has for the corresponding value of $q$ an isolated minimum. This is the *principle of Lagrange*. We shall now discuss the stability of equilibrium solutions which correspond with isolated maxima of the potential function $\phi(q)$. It is clear, that in this case the Hamiltonian is not a Lyapunov-function in the sense of theorem 8.1 or 8.3. Assume again that the critical point is $(p,q) = (0,0)$ and that $H(0,0) = 0$, so $\phi(0) = 0$. Also we assume that $\phi(q)$ can be expanded in a Taylor series to degree $2m + 1$ $(m \in I\!N)$ in a neighbourhood of $q = 0$ so that

$$(8.6) \qquad \phi(q) = -\sum_{i=1}^{n} a_i q_i^{2m} + O\|q\|^{2m+1} \text{ as } \|q\| \to 0,$$

with $a_i > 0$, $i = 1, \cdots, n$. If $m = 1$ we can use linearisation and theorem 7.3 to show that the trivial solution is unstable. If $m > 1$ we have to proceed in a different way.

We introduce the function

$$V(p, q) = \sum_{i=1}^{n} p_i q_i.$$

In each neighbourhood of $(0,0)$ the function $V$ takes positive (and negative) values. We compute the orbital derivative

$$
\begin{aligned}
L_t V(p, q) &= \sum_{i=1}^{n} (\dot{p}_i q_i + p_i \dot{q}_i) \\
&= \sum_{i=1}^{n} (-\frac{\partial \phi}{\partial q_i} q_i + p_i^2) \\
&= \sum_{i=1}^{n} (2m a_i q_i^{2m} + p_i^2) + O\|q\|^{2m+1}
\end{aligned}
$$

For the last step we used the expansion 8.6. We conclude that $L_t V$ is positively definite in a neighbourhood of $(0,0)$ whereas $V$ takes positive values. Using theorem 8.3 we conclude that $(0,0)$ is unstable.

We summarise as follows.

**Theorem 8.5.**

Consider the Hamilton function

$$H(p, q) = \frac{1}{2} \sum_{i=1}^{n} p_i^2 + \phi(q) , \ (p, q) \in I\!\!R^{2n}$$

with potential $\phi(q)$ which can be expanded in a Taylor series in a neighbourhood of each critical point. An isolated minimum of the potential corresponds with a stable equilibrium solution, an isolated maximum corresponds with an unstable equilibrium solution.

**Remark**

Of course one can weaken the assumption on the Taylor series of $\phi(q)$; the expansion 8.6 is often met in applications but it has been used only to facilitate the demonstration.

**Example 8.2.**

In example 2.11 we studied the equation

$$\ddot{x} + f(x) = 0$$

which can be viewed as the equation of motion corresponding to the Hamilton function

$$H(p, q) = \frac{1}{2} p^2 + \int^q f(\tau) d\tau.$$

To obtain the equation of motion we put $(p, q) = (\dot{x}, x)$. The function $\int^q f(\tau) d\tau$ can be identified with the potential $\phi(q)$ in theorem 8.6. The isolated minima and maxima of $\phi(q)$ correspond with respectively stable and unstable equilibrium solutions, see figure 8.3.

**Example 8.3**

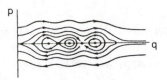

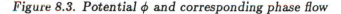

*Figure 8.3. Potential $\phi$ and corresponding phase flow*

A potential problem which plays an important part in the modern theory of Hamiltonian systems was formulated and studied by Hénon and Heiles. The Hamilton function is

$$H(p,q) = \frac{1}{2}(p_1^2 + p_2^2) + \phi(q_1, q_2)$$
$$= \frac{1}{2}(p_1^2 + p_2^2) + \frac{1}{2}(q_1^2 + q_2^2) + q_1^2 q_2 - \frac{1}{3}q_2^3.$$

The potential function $\phi(q)$ has the following critical points:
$(0,0)$ isolated minimum
$(0,1)$ and $(\pm\frac{1}{2}\sqrt{3}, -\frac{1}{2})$ are points where no maximum or minimum value is assumed.
Theorem 8.5, applied to this system, tells us that $(p_1, p_2, q_1, q_2) = (0,0,0,0)$ is a stable equilibrium solution. In a neighbourhood of the three critical points $(0,0,0,1)$ and $(0,0,\pm\frac{1}{2}\sqrt{3}, -\frac{1}{2})$ we perform linearisation after which we apply theorem 7.3 to conclude instability.

We shall now consider an example of a system with several first integrals. These integrals will be used to construct a Lyapunov function. In this construction we shall use the fact that a continuously differentiable function of a first integral is again a first integral. To see this consider the equation

$$\dot{x} = f(x) \text{ in } \mathbb{R}^n$$

with $k$ first integrals $I, \cdots, I_k$ $(k < n)$, so $L_t I_i(x) = 0$, $i = 1, \cdots, k$. The function $F : \mathbb{R}^k \to \mathbb{R}$ is continuously differentiable and $F(I, (x), \cdots, I_k(x))$ is a first integral of the equation:

$$L_t F = \sum_{i=1}^{k} \frac{\partial F}{\partial I_i} L_t I_i(x) = 0.$$

## Example 8.4

Consider a solid body rotating around a fixed point, which coincides with its centre of gravity. One can think for instance of a triaxial homogeneous ellipsoid which can rotate around its centre. $A$, $B$ and $C$ are the three moments of inertia; $x$, $y$ and $z$ are the projections of the angular velocity vector on the three coordinate axis. The equations of motion are

(8.7)
$$A\dot{x} = (B - C)yz$$
$$B\dot{y} = (C - A)zx$$
$$C\dot{z} = (A - B)xy$$

The axis corresponding with the moments of inertia concide with the coordinate axis. This means that a critical point represents an equilibrium solution corresponding with rotation of the solid body around one axis with a certain speed. We shall study the stability of the equilibrium solution $x = x_0 > 0$, $y = z = 0$ which corresponds with the smallest moment of inertia $A$ as we shall assume $0 < A < B < C$. The matrix arising after linearisation in $(x_0, 0, 0)$ is

$$\begin{pmatrix} 0 & 0 & 0 \\ 0 & 0 & \frac{C-A}{B}x_0 \\ 0 & \frac{A-B}{C}x_0 & 0 \end{pmatrix}$$

It is clear that this matrix has one eigen-value zero so that the Poincaré-Lyapunov theorems of chapter 7 do not apply. This is not unexpected as the critical points of system 8.7 are not isolated.

We can find first integrals of system 8.7 by deriving equations for $dx/dy$ and $dy/dz$ and by integrating these. We find the first integrals

$$V_1(x,y) = A(C - A)x^2 + B(C - B)y^2$$
$$V_2(y,z) = B(B - A)y^2 + C(C - A)z^2.$$

$V_1$, or better $V_1(x,y) - V_1(x_0, 0)$, is not sign definite in a neighbourhood of $(x_0, 0, 0)$, $V_2$ is semidefinite. One can construct a Lyapunov function in various ways, for instance

$$V(x,y,z) = [V_1(x,y) - V_1(x_0,0)]^2 + V_2(y,z).$$

The function $V$ is positively definite in a neighbourhood of $(x_0, 0, 0)$ while $L_t V = 0$. So the equilibrium solution $(x_0, 0, 0)$ is Lyapunov stable.

The reader should study the stability of the critical points which are found on the z-axis, corresponding with the largest moment of inertia $C$. With a similar computation one finds again stability.

## 8.4   Applications and examples

We mentioned already that, generally speaking, we do not know much about the existence and form of Lyapunov-functions. In each particular problem we shall have to use our intuition and to trust our luck. It is however easy to indicate the relation between the method of linearisation in chapter 7 and the direct method of Lyapunov.

To demonstrate this relation we shall not consider the general case, but we shall restrict ourselves to the system

$$\dot{x} = Ax + f(x)$$

with

$$\lim_{\|x\| \to 0} \frac{\|f(x)\|}{\|x\|} = 0, \ , \ x = (x_1, \cdots, x_n).$$

The constant $n \times n$-matrix $A$ has the diagonal form, the eigenvalues $\lambda_1, \cdots, \lambda_n$ are all real and negative. With the Poincaré-Lyapunov theorem 7.1 we conclude that the trivial solution $x = 0$ is asymptotically stable.

Consider on the other hand the function

$$V(x) = x_1^2 + x_2^2 + \cdots + x_n^2.$$

This function is positively definite. We compute the orbital derivative

$$
\begin{aligned}
L_t V &= 2 \sum_{i=1}^n x_i \dot{x}_i \\
&= 2 \sum_{i=1}^n \lambda_i x_i^2 + 2 \sum_{i=1}^n x_i f_i(x)
\end{aligned}
$$

For sufficiently small $\varepsilon > 0$ we have: if $\|x\| \le \delta(\varepsilon)$ then $\|f(x)\| \le \varepsilon \|x\|$ where $\delta \to 0$ if $\varepsilon \to 0$. As the eigenvalues $\lambda_i$ are all negative we find that $L_t V$ is negatively definite in a neighbourhood of $x = 0$. $V(x)$ is a Lyapunov-function and application of theorem 8.2 yields the asymptotic stability of $x = 0$.

We shall consider some other examples.

## Example 8.5

In a neighbourhood of $(0,0)$ we consider the system

$$
\begin{aligned}
\dot{x} &= a(t)y + b(t)x(x^2 + y^2) \\
\dot{y} &= -a(t)x + b(t)y(x^2 + y^2).
\end{aligned}
$$

The functions $a(t)$ and $b(t)$ are continuous for $t \ge t_0$. The trivial solution $(0,0)$ is stable if $b(t) \le 0$, unstable if $b(t) > 0$ for $t \ge t_0$. It is easy to demonstrate this; take

$$V(x,y) = x^2 + y^2.$$

as a Lyapunov-function. $V$ is positively definite whereas

$$L_t V = 2b(t)(x^2 + y^2)^2.$$

Application of the theorems 8.1 and 8.3 produces the required result.

The ideas which play a part in the proofs of Lyapunov's theorems can also be used to show the boundedness of solutions and attraction in a domain.

**Example 8.6**

Consider the equation for a nonlinear oscillator with linear damping

$$\ddot{x} + \mu\dot{x} + x + ax^2 + bx^3 = 0$$

with constants $\mu, a, b; \mu > 0$. The damping $\mu\dot{x}$ causes the trivial solution to be asymptotically stable, see also example 7.1. We introduce the energy of the non-linear oscillator without damping:

$$V(x,\dot{x}) = \frac{1}{2}\dot{x}^2 + \frac{1}{2}x^2 + \frac{1}{3}ax^3 + \frac{1}{4}bx^4.$$

We can find a neighbourhood $D$ of $(0,0)$, dependent in size on $a$ and $b$, in which $V$ is positively definite. Furthermore we have

$$
\begin{aligned}
L_t V &= \dot{x}\ddot{x} + x\dot{x} + ax^2\dot{x} + bx^3\dot{x} \\
&= -\mu\dot{x}^2.
\end{aligned}
$$

We can apply theorem 8.1 to find that $(0,0)$ is Lyapunov-stable but we cannot apply theorem 8.2 to obtain asymptotic stability as $L_t V \leq 0$. As $L_t V \leq 0$ in $D$, the solutions cannot leave $D$ (see the proof of theorem 8.1), also we have that the equality $L_t V = 0$ is only valid if $\dot{x} = 0$. However, if $x \neq 0$, $\dot{x} = 0$ is a transversal of the phase-flow so we conclude that $D$ is a domain of attraction of the trivial solution.

## 8.5 Exercises

8-1. Consider the two-dimensional system

$$
\begin{aligned}
\dot{x} &= -y + f(x,y) \\
\dot{y} &= \sin x
\end{aligned}
$$

The function $f(x,y)$ is smooth and nonlinear.
Give sufficient conditions for $f(x,y)$ so that $(0,0)$ is a stable equilibrium solution.

8-2. Equations of the form $\ddot{y} + p(t)\dot{y} + q(t)y = 0$ with $p(t), q(t)$ $C^1$ functions can be put in the form

$$\ddot{x} + w(t)x = 0 \quad , w(t) \text{ a } C^1 \text{ function,}$$

by the transformation of Liouville:

$$x(t) = y(t)e^{-\frac{1}{2}\int_{t_0}^{t} p(\tau)d\tau}.$$

Does Lyapunov-stability of $(x, \dot{x}) = (0,0)$ imply the Lyapunov- stability of $(y, \dot{y}) = (0,0)$?

8-3. Determine the stability of the solution $(x, \dot{x}) = (0,0)$ of the set of equations $\ddot{x} + x^n = 0$ $(n \in \mathbb{N})$.

8-4. Consider the system

$$\dot{x} = 2y(z-1) \ , \ \dot{y} = -x(z-1) \ , \ \dot{z} = xy.$$

   a. Show that the solution $(0,0,0)$ is stable

   b. Is this solution asymptotically stable?

8-5. Determine the stability of the solution $(x, y) = (0,0)$ of the system $\dot{x} = 2xy + x^3$ , $\dot{y} = x^2 - y^5$.

8-6. Consider the equation
$$\ddot{x} + \phi(t)x = 0$$

with $\phi \in C^1(\mathbb{R})$ , $\phi(t)$ is monotonic and

$$\lim_{t \to \infty} \phi(t) = c > 0.$$

   a. Can we apply theorem 6.2 to prove stability of $(x, \dot{x}) = (0,0)$. Consider also the counter-example 6.2.

   b. Prove that $(0,0)$ is stable.

8-7. The system $\dot{x} = Ax + f(x), x \in \mathbb{R}^n$ has the following properties. We have

$$\lim_{\|x\| \to 0} \frac{\|f(x)\|}{\|x\|} = 0$$

and the constant $n \times n$-matrix $A$ has $n$ distinct real eigenvalues $\lambda_1, \ldots, \lambda_n$ with $\lambda_1, \ldots, \lambda_p > 0$ and $\lambda_{p+1}, \ldots, \lambda_n < 0$, $0 < p < n$. Find a Lyapunov function to prove the instability of $x = 0$.

8-8. We are interested in conditions which quarantee that the following third-order oscillator

$$\dddot{x} + f(\dot{x})\ddot{x} + a\dot{x} + bx = 0$$

has a globally attracting trivial solution; a and b are positive constants.

8-9. Determine the stability of the trivial solution of

$$\dot{x} = xy^2 - \frac{1}{2}x^3 \ , \ \dot{y} = -\frac{1}{2}y^3 + \frac{1}{5}x^2y$$

# 9 Introduction to perturbation theory

This chapter is intended as an introduction for those readers who are not aquainted with the basics of perturbation theory. In that case it serves in preparing for the subsequent chapters.

## 9.1 Background and elementary examples

The development of the theory of the influence of small perturbations on solutions of differential equations started in the eighteenth century. Since Poincaré (around 1900) perturbation theory is flourishing again with important new fundamental insights and applications. In the eighteenth century one of the most important fields of application of perturbation theory is the theory describing the motion of celestial bodies. The motion of the earth around the sun according to Newton's laws envisaged as the dynamics of two bodies, point masses, bound by gravity, leads to periodic revolution in an elliptic orbit. However, with the accuracy of measurement acquired in those times one could establish considerable deviations of this motion in an elliptic orbit. Looking for causes of these deviations one considered the perturbational influence of the moon and the large planets like Jupiter and Saturn. Such models are leading to equations of motion which, apart from terms corresponding with the mutual attraction of the earth and the sun, contain small perturbation terms. The study of these new equations turns out to be very difficult, in fact the fundamental questions of this problem have not been solved up till now, but it started off the development of perturbation theory.

**Example 9.1**
A simple example of a perturbation arising in a natural way is the following problem. Consider a harmonic oscillation, described by the equation

(9.1) $$\ddot{x} + x = 0$$

In deriving this equation the effect of friction has been neglected; in practice however, friction will always be present. If the oscillator is such that the friction is small, an improved model for the oscillations is given by the equation

(9.2) $$\ddot{x} + \varepsilon \dot{x} + x = 0.$$

The term $\varepsilon \dot{x}$ is called "friction term" or "damping term" and this particular simple form of the friction term has been based on certain assumptions concerning the mechanics of friction. The parameter $\varepsilon$ is small:

$$0 \leq \varepsilon << 1$$

as will always be the case, in this chapter and in subsequent chapters.
If one puts $\varepsilon = 0$ in equation 9.2 one recovers the original equation 9.1; we call equation 9.1 the "*unperturbed problem*". We shall always proceed while assuming that we have sufficient knowledge of the solutions of the unperturbed problem. If this is not the case, it does not make much sense, generally speaking, to study the more complicated problem with $\varepsilon > 0$.

One of the interesting conclusions which we shall draw is, that in a great many problems, the introduction of small perturbations triggers off qualitatively and quantitatively behaviour of the solutions which diverges very much from the behaviour of the solutions of the unperturbed problem. One can observe this already in the simple example of the damped oscillator described by equation 9.2. We shall consider some other examples.

**Example 9.2**

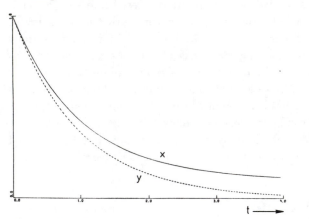

*Figure 9.1.* $0 \leq x \leq 1$

Consider the initial value problem

$$\dot{x} = -x + \varepsilon \, , \; x(0) = 1.$$

The solution is $x(t) = \varepsilon + (1 - \varepsilon)e^{-t}$. The unperturbed problem is

$$\dot{y} = -y \, , \; y(0) = 1.$$

The solution is $y(t) = e^{-t}$. It is clear that

$$| \, x(t) - y(t) \, | = \varepsilon - \varepsilon e^{-t} \leq \varepsilon \, , \; t \geq 0.$$

The error, arising in approximating $x(t)$ by $y(t)$ is never bigger that $\varepsilon$.

**Example 9.3**

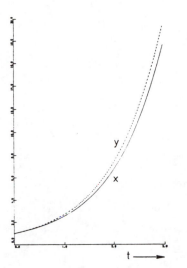

*Figure 9.2.* $0 \le x \le 20$

The situation is very different for the problem

$$\dot{x} = +x + \varepsilon \, , \; x(0) = 1.$$

The solution is $x(t) = -\varepsilon + (1 + \varepsilon)e^t$. The unperturbed problem is

$$\dot{y} = +y \, , \; y(0) = 1$$

with solution $y(t) = e^t$. We find

$$\mid x(t) - y(t) \mid = \varepsilon \mid 1 - e^t \mid .$$

On the interval $0 \le t \le 1$, the error caused by approximating $x(t)$ by $y(t)$ is of the order of magnitude $\varepsilon$. This is not true anymore for $t \ge 0$, where the difference increases without bounds.

**Example 9.4**

In example 9.3 the solutions are not bounded for $t \ge 0$. We shall show now that also in the case of bounded solutions the difference between the solutions of the perturbed and the unperturbed problem can be considerable. Consider the initial value problem

$$\ddot{x} + (1 + \varepsilon)^2 x = 0 \, , \; x(0) = 0, \; \dot{x}(0) = 1.$$

The solution is

$$x(t) = \frac{1}{1 + \varepsilon} \sin(1 + \varepsilon)t.$$

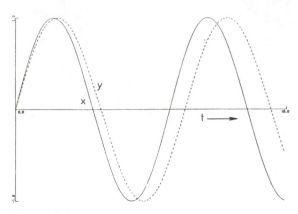

*Figure 9.3*

The unperturbed problem is

$$\ddot{y} + y = 0 \ , \ y(0) = 0 \ , \ \dot{y}(0) = 1$$

with solution $y(t) = \sin t$. The reader should analyse the difference $x(t) - y(t)$. The solutions are close for a long time, but if one waits long enough, take for example $t = \pi/(2\varepsilon)$, the difference becomes considerable.

These examples do illustrate that in constructing approximations of solutions of initial value problems, one has to indicate on what interval of time one is looking for an approximation. In many cases we would like this interval of time to be as large as possible.
We shall introduce now a number of concepts which enable us to estimate vector functions in terms of a small parameter $\varepsilon$.

## 9.2   Basic material

Consider the vector function $f : R \times R^n \times R \to R^n$; the function $f(t, x, \varepsilon)$ is continuous in the variables $t \in R$ and $x \in D \subset R^n$, $\varepsilon$ is a small parameter. The function $f$ has to be expanded with respect to the small parameter $\varepsilon$. In the simple case that $f$ has a Taylor expansion with respect to $\varepsilon$ near $\varepsilon = 0$ we have

$$f(t, x, \varepsilon) = f(t, x, 0) + \varepsilon f_1(t, x) + \varepsilon^2 f_2(t, x) + \cdots \varepsilon^n f_n(t, x) + \cdots$$

with coefficients $f_1, f_2, \cdots$ which depend on $t$ and $x$. The expressions $\varepsilon, \varepsilon^2, \cdots, \varepsilon^n, \cdots$ are called order functions.
More often than not we do not know the vector function $f$ explicitly, for instance as it has been defined implicitly as the solution of an equation with (initial) conditions. In a number of cases we also do not know apriori which form the expansion

with respect to $\varepsilon$ will take. Very often we shall look for an expansion of the form

$$f(t, x, \varepsilon) = \sum_{n=0}^{N} \delta_n(\varepsilon) f_n(t, x) + \cdots$$

in which $\delta_n(\varepsilon), n = 0, 1, 2, \cdots$ are order functions. We shall make this more precise.

### Definition
The function $\delta(\varepsilon)$ is continuous and positive in $(0, \varepsilon_0]$ and it has the property that $\lim_{\varepsilon \to 0} \delta(\varepsilon)$ exists whereas $\delta(\varepsilon)$ is monotonically decreasing as $\varepsilon$ tends to zero; $\delta(\varepsilon)$ is called an order function.

In the case that $f$ has a Taylor expansion, the order functions which have been used are

$$\{\varepsilon^n\}_{n=0}^{\infty}$$

Other examples of order functions on $(0, 1]$ are

$$\varepsilon \mid \ln\varepsilon \mid, \ \sin\varepsilon, \ 2, \ e^{-1/\varepsilon}$$

Very often we shall compare the magnitude of order functions with a characterisation like "$\varepsilon^4$ goes faster to zero that $\varepsilon^{2}$". What we mean is that we compare the behaviour of these two order functions as $\varepsilon$ tends to zero; for this comparison we shall use the Landau O-symbols.

### Definition

a. $\delta_1(\varepsilon) = O(\delta_2(\varepsilon))$ as $\varepsilon \to 0$ if there exists a constant $k$ such that $\delta_1(\varepsilon \leq k\delta_2(\varepsilon)$ as $\varepsilon \to 0$.

b. $\delta_1(\varepsilon) = o(\delta_2(\varepsilon))$ as $\varepsilon \to 0$ if

$$\lim_{\varepsilon \to 0} \frac{\delta_1(\varepsilon)}{\delta_2(\varepsilon)} = 0.$$

### Remark
In a number of cases the following limit exists

$$\lim_{\varepsilon \to 0} \frac{\delta_1(\varepsilon}{\delta_2(\varepsilon)} = k.$$

In this case we find with the definition $\delta_1(\varepsilon) = O(\delta_2(\varepsilon))$.
So if this limit exists, we have in a simple way the $O$-estimate for order functions.
Examples are

$$\varepsilon^4 = 0(\varepsilon^2) \text{ as } \varepsilon \to 0, \sin\varepsilon = O(\varepsilon) \text{ as } \varepsilon \to 0, \varepsilon \mid \ln\varepsilon \mid = 0(1) \text{ as } \varepsilon \to 0.$$

In perturbation theory we usually omit "as $\varepsilon \to 0$" as this is always the case to be considered.

We are now able to compare functions of $\varepsilon$, but this does not apply to vector functions $f(t, x, \varepsilon)$. Consider for instance the function

$$\varepsilon t \sin x , \; 0 \le t \le 1 , \; x \in \mathbf{R}.$$

Intuitively we would estimate $\varepsilon t \sin x = O(\varepsilon)$ on the domain $[0, 1] \times \mathbf{R}$; to do this we have to extend our definitions.

## Definition

Consider the vector function $f(t, x, \varepsilon)$, $t \in I \subset \mathbf{R}$, $x \in D \subset \mathbf{R}^n$, $0 \le \varepsilon \le \varepsilon_0$;

a. $f(t, x, \varepsilon)$ is $O(\delta(\varepsilon))$ if there exists a constant $k$ such that $\|f\| \le k\delta(\varepsilon)$ as $\varepsilon \to 0$ with $\delta(\varepsilon)$ an order function ($\|.\|$ is the sup norm on $I \times D$, see section 1.1).

b. $f(t, x, \varepsilon)$ is $o(\delta(\varepsilon))$ as $\varepsilon \to 0$ if $\lim\limits_{\varepsilon \to 0} \dfrac{\|f\|}{\delta(\varepsilon)} = 0$.

In the example given above we have now

$$\varepsilon t \sin x = O(\varepsilon) , \; 0 \le t \le 1 , \; x \in \mathbf{R}.$$

Note however that the $O(\varepsilon)$ estimate does not hold for the functions

$$\varepsilon t \sin x \quad , t \ge 0 \quad , x \in \mathbf{R}$$
$$\varepsilon^2 t \sin x \quad , t \ge 0 \quad , x \in \mathbf{R}.$$

As we are especially interested in initial value problems for differential equations, the variable $t$ plays a special part. Often we shall study a solution or its approximation on an interval which we would like to take as large as possible. So in these cases we shall not fix the interval of time apriori.

It turns out to be useful to characterise the size of the interval of time in terms of the small parameter $\varepsilon$.

## Definition time-scale

Consider the vector function $f(t, x, \varepsilon)$, $t \ge 0$, $x \in D \subset \mathbf{R}^n$, and the order functions $\delta_1(\varepsilon)$ and $\delta_2(\varepsilon)$; $f(t, x, \varepsilon) = O(\delta_1(\varepsilon))$ as $\varepsilon \to 0$ on the time-scale $1/\delta_2(\varepsilon)$ if the estimate is valid for $x \in D$, $0 \le \delta_2(\varepsilon)t \le C$ with $C$ a constant independent of $\varepsilon$. The formulation in the case of the $o$-symbol runs in an analogous way.

## Example 9.5

$f_1(t, x, \varepsilon) = \varepsilon t \sin x$, $t \ge 0$, $x \in \mathbf{R}$;

$f_1(t, x, \varepsilon) = O(\varepsilon)$ on the time-scale 1, $f_1(t, x, \varepsilon) = O(1)$ on the time-scale $1/\varepsilon$.

$f_2(t, x, \varepsilon) = \varepsilon^2 t \sin x$ , $t \geq 0$, $x \in \mathbb{R}$;
$f_2(t, x, \varepsilon) = O(\varepsilon)$ on the time-scale $1/\varepsilon$,
$f_2(t, x, \varepsilon) = O(\varepsilon^{1/2})$ on the time-scale $1/\varepsilon^{3/2}$.

We are now able to introduce the concept of an asymptotic approximation of a function.

**Definition**
Consider the functions $f(t, x, \varepsilon)$, $g(t, x, \varepsilon)$, $t \geq 0$, $x \in D \subset \mathbb{R}^n$; $g(t, x, \varepsilon)$ is an asymptotic approximation of $f(t, x, \varepsilon)$ on the time-scale $1/\delta(\varepsilon)$ if

$$f(t, x, \varepsilon) - g(t, x, \varepsilon) = o(1) \text{ as } \varepsilon \to 0 \text{ on the time-scale } 1/\delta(\varepsilon).$$

Approximation of functions will take the form of asymptotic expansions like

$$f(t, x, \varepsilon) = \sum_{n=0}^{N} \delta_n(\varepsilon) f_n(t, x, \varepsilon) + o\left(\delta_N(\varepsilon)\right)$$

with $\delta_n(\varepsilon), n = 0, \cdots, N$, order functions such that $\delta_{n+1}(\varepsilon) = o\left(\delta_n(\varepsilon)\right)$, $n = 0, \cdots, N-1$. The expansion coefficients $f_n(t, x, \varepsilon)$ are bounded and $O(1)$ as $\varepsilon \to 0$ on the given $t, x$- domain.

**Example 9.6**
$f_1(t, x, \varepsilon) = x \sin(t + \varepsilon t)$, $t \geq 0, 0 \leq x \leq 1$.
The function $x \sin t$ is an asymptotic approximation of $f_1(t, x, \varepsilon)$ on the time-scale 1.
$f_2(t, x, \varepsilon) = \sin \varepsilon + \varepsilon^3 t e^{-\varepsilon t}, t \geq 0$.
The function $\varepsilon$ is an asymptotic approximation of $f_2(t, x, \varepsilon)$ on the time-scale $1/\varepsilon$.

## 9.3  Naïve expansion

Consider the initial value problem

$$\dot{x} = f(t, x, \varepsilon) , \ x(0) \text{ given}$$

$t \geq 0, x \in D \subset \mathbb{R}^n$. If we can expend $f(t, x, \varepsilon)$ in a Taylor series with respect to $\varepsilon$

$$f(t, x, \varepsilon) = f_0(t, x) + \varepsilon f_1(t, x) + \varepsilon^2 \cdots$$

then one might suppose that a similar expansion exists for the solution

$$x(t) = x_0(t) + \varepsilon x_1(t) + \varepsilon^2 \cdots$$

Substitution of the expansion in the equation, equating coefficients of corresponding powers of $\varepsilon$ and subsequently imposing the initial values produces a so-called formal expansion of the solution. It seems natural to expect that the formal expansion represents an asymptotic approximation of the solution. We shall show that this is correct, but only in a very restricted sense; first a simple example.

## Example 9.7
Consider the oscillator with damping

$$(9.3) \qquad \ddot{x} + 2\varepsilon\dot{x} + x = 0$$

with initial values $x(0) = a$, $\dot{x}(0) = 0$. The solution of the initial value problem is

$$(9.4) \qquad x(t) = ae^{-\varepsilon t}\cos(\sqrt{1 - \varepsilon^2}t) + \varepsilon\frac{a}{\sqrt{1 - \varepsilon^2}}e^{-\varepsilon t}\sin(\sqrt{1 - \varepsilon^2}t)$$

Ignoring the explicit solution we substitute

$$x(t) = x_0(t) + \varepsilon x_1(t) + \varepsilon^2 \cdots$$

Equating coefficients of equal powers of $\varepsilon$, we find

$$\ddot{x}_0 + x_0 = 0$$
$$\ddot{x}_n + x_n = -2\dot{x}_{n-1}, \; n = 1, 2, \cdots$$

The initial values do not depend on $\varepsilon$ so it is natural to put

$$x_0(0) = a \; , \quad \dot{x}_0(0) = 0$$
$$x_n(0) = 0 \; , \quad \dot{x}_n(0) = 0, n = 1, 2, \cdots$$

Solving the equations and applying the initial conditions we obtain

$$x_0(t) = a\cos t$$
$$x_1(t) = a\sin t - at\cos t, \text{ etc.}$$

The formal expansion is

$$x(t) = a\cos t + a\varepsilon(\sin t - t\cos t) + \varepsilon^2 \cdots$$

This expansion corresponds with an oscillation with *increasing* amplitude; it is easy to see that on the time-scale $1/\varepsilon$ the formal expansion does not represent an asymptotic approximation of the solution (substitute for instance $t = 1/\varepsilon$ in the solution and in the formal expansion).

Formal expansion as in example 9.7 can be characterised as follows.

## Theorem 9.1
Consider the initial value problem

$$\dot{x} = f_0(t, x) + \varepsilon f_1(t, x) + \cdots + \varepsilon^m f_m(t, x) + \varepsilon^{m+1}R(t, x, \varepsilon)$$

with $x(t_0) = \eta$ and $|t - t_0| \leq h$, $x \in D \subset \mathbb{R}^n$, $0 \leq \varepsilon \leq \varepsilon_0$. Assume that in this domain we have

a. $f_i(t, x), i = 0, \cdots, m$ continuous in $t$ and $x$, $(m + 1 - i)$ times continuously differentiable in $x$;

b. $R(t, x, \varepsilon)$ continuous in $t, x$ and $\varepsilon$, Lipschitz-continuous in $x$.

Substitution in the equation for $x$ the formal expansion

$$x_0(t) + \varepsilon x_1(t) + \cdots + \varepsilon^m x_m(t),$$

Taylor expansion with respect to powers of $\varepsilon$, equating corresponding coefficients and applying the initial values $x_0(t_0) = \eta$, $x_i(t) = 0$, $i = 1, \cdots, m$ produces an approximation of $x(t)$:

$$\|x(t) - (x_0(t) + \varepsilon x_1(t) + \cdots + \varepsilon^m x_m(t))\| = O(\varepsilon^{m+1})$$

on the time-scale 1.

**Proof**

Substitution of the formal expansion in the equation produces

$$\dot{x}_0 + \varepsilon \dot{x}_1 + \cdots = f_0(t, x_0 + \varepsilon x_1 + \cdots) + \varepsilon f_1(t, x_0 + \varepsilon x_1 + \cdots) + \cdots$$

where the dots indicate terms starting with $\varepsilon^2$, $\varepsilon^3$ etc. Expansion in a Taylor series and equating coefficients of equal powers of $\varepsilon$ yields

$$\dot{x}_0 = f_0(t, x_0) \text{ (unperturbed equation)}$$
$$\dot{x}_1 = f_1(t, x_0) + \frac{\partial f_0}{\partial x}(t, x_0) x_1.$$

The equation for $x_i$, $i = 1, 2, \cdots$ is of the form

$$\dot{x}_i = A_i(t) + B_i(t) x_i,$$

in which $A_i(t)$ and $B_i(t)$ depend on $x_0(t), \cdots, x_{i-1}(t)$, i.e. the unperturbed equation is nonlinear, the subsequent equations are linear with variable coefficients. The equivalent integral equation for the initial value problem is

$$x(t) = \eta + \int_{t_0}^{t} f_0(\tau, x(\tau)) d\tau + \cdots + \varepsilon^m \int_{t_0}^{t} f_m(\tau, x(\tau)) d\tau +$$

$$+ \varepsilon^{m+1} \int_{t_0}^{t} R(\tau, x(\tau), \varepsilon) d\tau.$$

We start by choosing $m = 0$; we have

$$x_0(t) = \eta + \int_{t_0}^{t} f_0(\tau, x_0(\tau)) d\tau$$

and

$$x(t) - x_0(t) = \int_{t_0}^{t} [f_0(\tau, x(\tau)) - f_0(\tau, x_0(\tau))] d\tau +$$

$$\varepsilon \int_{t_0}^{t} R(\tau, x(\tau), \varepsilon) d\tau.$$

Using the Lipschitz-continuity of $f_0$, Lipschitz-constant $L$, and the boundedness of $R$, $\|R\| \leq M$, we find for $t \geq t_0$

$$\|x(t) - x_0(t)\| \leq \int_{t_0}^{t} L\|x(\tau) - x_0(\tau)\|d\tau + \varepsilon M(t - t_0).$$

Application of theorem 3.1 (Gronwall) with $\delta_1 = L$, $\delta_2 = \varepsilon M$ and $\delta_3 = 0$ produces

$$\|x(t) - x_0(t))\| \leq \varepsilon \frac{M}{L} e^{L(t-t_0)} - \varepsilon \frac{M}{L}.$$

We conclude that $x(t) - x_0(t) = O(\varepsilon)$ on the time-scale 1.
It follows that we may put

$$x(t) = x_0(t) + \varepsilon\phi(t, \varepsilon)$$

with $\phi(t, \varepsilon)$ bounded by a constant independent of $\varepsilon$ on the time-scale 1.
Now we choose $m = 1$ and the reader can check that we may use the same technique, integral equations and Gronwall, to prove that $\phi(t, \varepsilon) - x_1(t) = O(\varepsilon)$ on the time-scale 1. It is clear that, continuing in this way, the theorem follows. □

The expansion which we have obtained is an asymptotic approximation of the solution. A natural question is whether, on taking the limit $m \to \infty$, while assuming that the righthand side of the equation can be represented by a convergent power series with respect to $\varepsilon$, we obtain in this way a convergent series for $x(t)$. This question has been answered in a positive way by Poincaré.

## 9.4   The Poincaré expansion theorem

Consider the initial value problem

(9.5) $$\dot{x} = f(t, x, \varepsilon) , \ x(t_0) = \eta,$$

where we assume that $f(t, x, \varepsilon)$ can be expanded in a convergent Taylor series with respect to $\varepsilon$ and $x$ in a certain domain. The unperturbed problem is

$$\dot{x}_0 = f(t, x_0, 0)$$

In many applications, for instance if we are looking for periodic solutions, we do not know precise initial conditions; we allow for this by admitting deviations of order $\mu$

$$x(t_0) = x_0(t_0) + \mu$$

with $\mu$ a constant, at this point independent of $\varepsilon$. We translate

$$x = y + x_0(t)$$

to find for $y$

(9.6)
$$\dot{y} = F(t, y, \varepsilon) \, , \, y(t_0) = \mu$$

with $F(t, y, \varepsilon) = f(t, y + x_0(t), \varepsilon) - f(t, x_0(t), 0)$.

The expansion properties of $f(t, x, \varepsilon)$ imply that $F(t, y, \varepsilon)$ possesses a convergent power expansion with respect to $y$ and $\varepsilon$ in a neighbourhood of $y = 0, \, \varepsilon = 0$.

The proof of theorem 9.1, in making use of the Lipschitz-continuity of the derivatives to order $m$, can be viewed as contraction in a $C^m$-space. In the subsequent proof we use contraction in a Banach space of analytic functions.

## Theorem 9.2 (Poincaré)

Consider the initial value problem 9.6

$$\dot{y} = F(t, y, \varepsilon) \, , \, y(t_0) = \mu$$

with $|t - t_0| \leq h, y \in D \subset \mathbb{R}^n, 0 \leq \varepsilon \leq \varepsilon_0, 0 \leq \mu \leq \mu_0$. If $F(t, y, \varepsilon)$ is continuous with respect to $t, y$ and $\varepsilon$ and can be expanded in a convergent power series with respect to $y$ and $\varepsilon$ for $\|y\| \leq \rho, \, 0 \leq \varepsilon \leq \varepsilon_0$, then $y(t)$ can be expanded in a convergent power series with respect to $\varepsilon$ and $\mu$ in a neighbourhood of $\varepsilon = \mu = 0$, convergent on the time-scale 1.

## Proof

Starting with the function $y^{(0)}(t) = \mu$ the usual contraction process produces a sequence

$$y^{(n+1)}(t) = \mu + \int_{t_0}^{t} F(\tau, y^{(n)}(\tau), \varepsilon) d\tau \, , \, n = 0, 1, 2, \cdots$$

which converges towards the solution of the initial value problem (see for instance Walter (1976) section 6).

The sequence is uniformly convergent for $|t - t_0| \leq h$. We shall use now the expansion properties of $F(t, y, \varepsilon)$ to develop at each iteration step $y^{(n)}(t)$ with respect to $\varepsilon$ and $\mu$. Integration of the powerseries for $y^{(n)}(t)$ yields a power series for $y^{(n+1)}(t)$ which is convergent in the same domain.

We are intending now to use the theorem of (complex) analytic function theory which tells us that the limiting function of a uniformly convergent series of analytic functions is also analytic (Grauert and Fritzsche (1976), theorem 3.5 in chapter 1.3). To apply this theorem we perform the analytical continuation of the functions $\{y^{(n)}(t)\}_{n=0}^{\infty}$ in the parameters $\varepsilon$ and $\mu$ on a domain $D \subset \mathbb{C}^2$ with positive measure; the theorem uses integration in $D$ for a certain Cauchy integral, in which $t$ plays the part of a real parameter. So the limiting function $y(t)$ is analytic in $\varepsilon$ and $\mu$. Finally we take a closer look at the interval of time on which the expansion is valid. On the domain of definition of $F$ we have

$$\|F\|_{\text{sup}} = M.$$

The expansion properties which we have used, require that the solution remains in a $\rho$-neighbourhood of $y = \mu$ at each iteration. For $n = 1$ we have

$$\|y^{(1)} - \mu\| \le \rho$$

so that

$$\|\mu\| + M|t - t_0| \le \rho.$$

It is easy to see that on requiring

$$\|y^{(n)} - \mu\| \le \rho$$

we have again

$$\|\mu\| + M|t - t_0| \le \rho$$

so that

$$|t - t_0| \le \rho/M - \|\mu\|/M.$$

So the expansion of the solution holds for $|t - t_0| \le \min(h, \frac{\rho - \|\mu\|}{M})$. If $\|\mu\| \ll \rho$ and $h$ is not $o(1)$, the expansion is clearly valid on the time-scale 1.

**Remark**

In the theorem we have assumed that $F$ (or $f$) depends on one parameter $\varepsilon$. It is easy to extend the theorem to the case of an arbitrary finite number of parameters.

## 9.5   Exercises

9-1. Approximate the periodic solution of the equation

$$\ddot{x} + x - \varepsilon x^3 = 0$$

starting in $x(0) = 1, \dot{x}(0) = 0$.

9-2. In exercise 7.4 we considered two competing species; now we choose the interaction to be small by putting $a = \varepsilon$:

$$\begin{aligned}
\dot{x} &= x - x^2 - \varepsilon xy \\
\dot{y} &= y - y^2 - \varepsilon xy.
\end{aligned}$$

a. Starting with positive initial values $x(0) = x_0$, $y(0) = y_0$, compute an expansion of the solution up till $O(\varepsilon)$.

b. We know that $\lim_{t \to \infty}(x(t), y(t)) = (\frac{1}{1+\varepsilon}, \frac{1}{1+\varepsilon})$ (attracting node). What is the limiting behaviour as $t \to \infty$ of the approximation? May we conclude that in this case the approximation is valid for all time?

9-3. We consider initial value problems satisfying the conditions of theorem 9.1.

    a. $\quad \dot{x} = \varepsilon f_1(t, x) + \varepsilon^2 f_2(t, x), \quad x(t_0) = \eta$
          $\dot{y} = \varepsilon f_1(t, y) , \quad y(t_0) = \eta.$
       Give the error estimate $\|x(t) - y(t)\|$ according to theorem 9.1.
       Can we improve upon the time-scale?

    b. The same questions for the problems

$$\dot{x} = \varepsilon^2 f_2(t, x) + \varepsilon^3 f_3(t, x) , \quad x(t_0) = \eta$$
$$\dot{y} = \varepsilon^2 f_2(t, y), \quad y(t_0) = \eta.$$

9-4. We are interested in the existence of periodic solutions of the perturbed Hamiltonian system

$$\ddot{x} + x - \lambda x^2 = \varepsilon(1 - x^2)\dot{x} , \quad \lambda > 0.$$

    a. Determine the location of the periodic solutions of the unperturbed problem ($\varepsilon = 0$).

    b. Characterise the critical points if $0 < \varepsilon \ll 1$.

    c. Determine with a certain accuracy the domain where periodic solutions might be found.

    d. Apply the Bendixson criterion (theorem 4.1) to the problem; this produces an upperbound for $\lambda$.

# 10 The Poincaré-Lindstedt method

In this chapter we shall show how to find convergent series approximations of *periodic* solutions by using the expansion theorem and the periodicity of the solution. This method is usually called after Poincaré and Lindstedt, it is also called the continuation method.

## 10.1 Periodic solutions of autonomous second-order equations

In this section we shall consider equations of the form

(10.1) $$\ddot{x} + x = \varepsilon f(x, \dot{x}, \varepsilon), \quad (x, \dot{x}) \in D \subset \mathbf{R}^2.$$

If $\varepsilon = 0$, all nontrivial solutions are $2\pi$-periodic. To start with we assume that periodic solutions of equation 10.1 exist for small, positive values of $\varepsilon$. See chapter 4 for theorems and examples; in section 10.4 we shall return to the question of existence.

Furthermore we shall assume that in $D$ and for $0 \leq \varepsilon \leq \varepsilon_0$, the requirements of the Poincaré expansion theorem (section 9.4) have been satisfied. Both the period $T$ of the solution and the initial values (and so the location of the solution in the phase-plane) will depend on the small parameter $\varepsilon$. This is the reason to put

$$T = T(\varepsilon) , \quad x(0) = a(\varepsilon) , \quad \dot{x}(0) = 0.$$

Note that putting $\dot{x} = 0$ for periodic solutions of equation 10.1 is no restriction. The approximations which we shall construct concern solutions which exist for all time. The time-scale of validity of the approximations will therefore play a part in the discussion. The expansion theorem tells us that, on the time-scale 1,

$$\lim_{\varepsilon \to 0} x(t, \varepsilon) = a(0) \cos t$$

with $a(0)$ at this point unknown; moreover we have clearly $T(0) = 2\pi$. In the sequel we shall expand both the initial value and the period with respect to the small parameter $\varepsilon$. It is convenient to transform $t \to \theta$ such that, in the new time-like variable $\theta$, the periodic solution is $2\pi$-periodic. We transform

$$wt = \theta , \quad w^{-2} = 1 - \varepsilon \eta(\varepsilon).$$

Equation 10.1 becomes after transforming and with notation $x' = dx/d\theta$

(10.2) $$x'' + x = \varepsilon[\eta x + (1 - \varepsilon \eta) f(x, (1 - \varepsilon \eta)^{-\frac{1}{2}} x', \varepsilon)],$$

$$\text{initial values } x(0) = a(\varepsilon), x'(0) = 0.$$

We shall abbreviate equation 10.2 by

$$x'' + x = \varepsilon g(x, x', \varepsilon, \eta).$$

The initial value $a$ and the parameter $\eta$, which determines the unknown period, have to be chosen such that we obtain a $2\pi$-periodic solution in $\theta$ of equation 10.2. By variation of constants we find that the initial value problem for equation 10.2 is equivalent with the following integral equation

$$x(\theta) = a\cos\theta + \varepsilon \int_0^\theta \sin(\theta - \tau)g(x(\tau), x'(\tau), \varepsilon, \eta)d\tau$$

For the periodic solution we have $x(\theta) = x(\theta + 2\pi)$ which yields the periodicity condition

$$\int_\theta^{\theta + 2\pi} \sin(\theta - \tau)g(x(\tau), x'(\tau), \varepsilon, \eta)d\tau = 0.$$

Expanding $\sin(\theta - \tau)$ we find two independent conditions

(10.3)
$$\int_0^{2\pi} \sin\tau g(x(\tau), x'(\tau), \varepsilon, \eta)d\tau = 0$$
$$\int_0^{2\pi} \cos\tau g(x(\tau), x'(\tau), \varepsilon, \eta)d\tau = 0.$$

The periodic solution depends on $\varepsilon$ but also on $a$ and $\eta$, so system 10.3 can be viewed as a system of two equations with two unknowns, $a$ and $\eta$. According to the implicit function theorem this system of equations 10.3 is uniquely solvable in a neighbourhood of $\varepsilon = 0$ if the corresponding Jacobian does not vanish. Writing system 10.3 in the form $F_1(a, \eta) = 0$, $F_2(a, \eta) = 0$ this means that in a neighbourhood of $\varepsilon = 0$

(10.4)
$$\frac{\partial(F_1, F_2)}{\partial(a, \eta)} \neq 0.$$

If condition 10.4 has been satisfied, we have with the assumptions on the righthand-side of equation 10.1, that $a(\varepsilon)$ and $\eta(\varepsilon)$ can be expanded in a Taylorseries with respect to $\varepsilon$. From equation 10.2 we find for system 10.3 with $\varepsilon = 0$

$$\int_0^{2\pi} \sin\tau f(a(0)\cos\tau, -a(0)\sin\tau, 0)d\tau = 0$$

(10.5)

$$\pi\eta(0)a(0) + \int_0^{2\pi} \cos\tau f(a(0)\cos\tau, -a(0)\sin\tau, 0)d\tau = 0.$$

Applying 10.4 to system 10.5 we find the condition (notation : $f = f(x, y, \varepsilon)$)

(10.6)
$$a(0)\int_0^{2\pi} [\frac{1}{2}\sin 2\tau \frac{\partial f}{\partial x}(a(0)\cos\tau, -a(0)\sin\tau, 0) +$$
$$- \sin^2\tau \frac{\partial f}{\partial y}(a(0)\cos\tau, -a(0)\sin\tau, 0)]d\tau \neq 0.$$

## Remark

If $\varepsilon = 0$, all solutions of equation 10.1 are periodic. Condition 10.4 and consequently condition 10.6, is a condition for the existence of an isolated periodic solution which branches off for $\varepsilon > 0$; see also the discussion in section 10.4. If, however, there exists for $\varepsilon > 0$ a continuous family of periodic solutions, a one-parameter family depending on $a(\varepsilon)$, then condition 10.4 or 10.6 will not be satisfied. If we know apriori that this family of periodic solutions exists, then we can of course still apply the Poincaré-Lindstedt method. In this respect the reader should study the equation

$$\ddot{x} + x = \varepsilon f(x).$$

See also the examples 2.11 and 2.14.

We conclude that if condition 10.4 has been satisfied, the corresponding periodic solution of equation 10.2 can be represented by the convergent series

(10.7)
$$x(\theta) = a(0) \cos \theta + \sum_{n=1}^{\infty} \varepsilon^n \gamma_n(\theta)$$

in which $\theta = \omega t = (1 - \varepsilon \eta)^{-\frac{1}{2}} t$ while using the convergent series

$$a = \sum_{n=0}^{\infty} \varepsilon^n a_n \, , \, \eta = \sum_{n=0}^{\infty} \varepsilon^n \eta_n \, , \, x(0) = a(0) + \sum_{n=1}^{\infty} \varepsilon^n \gamma_n(0), \, 0 < \varepsilon \leq \varepsilon_0.$$

The expansions of $T(\varepsilon)$ and $a(\varepsilon)$ which we used in the beginning of the construction have been validated in this way.

We shall now demonstrate the construction of the Poincaré-Lindstedt method for an important example.

## Example (van der Pol equation) 10.1

We showed in section 4.4 that the equation

$$\ddot{x} + x = \varepsilon(1 - x^2)\dot{x}$$

has one periodic solution for all positive values of $\varepsilon$. We take $\varepsilon$ small and we put again

$$\omega t = \theta, \omega^{-2} = 1 - \varepsilon \eta(\varepsilon)$$

so that the van der Pol equation transforms to

(10.8)
$$x'' + x = \varepsilon[\eta x + (1 - \varepsilon \eta)^{1/2}(1 - x^2)x']$$

with $x(0) = a(\varepsilon)$, $x'(0) = 0$. We can determine $a(0)$ and $\eta(0)$ with the system of equations 10.5. We find

$$\tfrac{1}{2}a(0)(1 - \tfrac{1}{4}a(0)^2) = 0$$
$$\eta(0)a(0) = 0$$

with non-trivial solutions $a(0) = 2$, $\eta(0) = 0$ (putting $a(0) = -2$ doesnot yield a new periodic solution). Condition 10.4 has been satisfied. To determine the terms of the series 10.7 systematically, we substitute the series with its derivatives into equation 10.8. Collecting terms which are coefficients of equal powers of $\varepsilon$ produces equations for the coefficients $\gamma_n(\theta)$. Then we apply the periodicity condition; this is aequivalent to using system 10.3 where $g$ has been expanded with respect to powers of $\varepsilon$. We write

$$x(\theta) = a_0 \cos\theta + \varepsilon\gamma_1(\theta) + \varepsilon^2\gamma_2(\theta) + \cdots,$$
$$x'(\theta) = -a_0 \sin\theta + \varepsilon\gamma_1'(\theta) + \varepsilon^2\gamma_2'(\theta) + \cdots \text{ etc.}$$

to find after substitution

(10.9)
$$\gamma_1'' + \gamma_1 = a_0\eta_0 \cos\theta + a_0(\tfrac{1}{4}a_0^2 - 1)\sin\theta +$$

$$\frac{1}{4}a_0^3 \sin 3\theta.$$

The solutions of equation 10.9 are periodic if $a_0 = 2$, $\eta_0 = 0$. The general solution of the equation is

$$\gamma_1(\theta) = A_1 \cos\theta + B_1 \sin\theta - \frac{a_0^3}{32}\sin 3\theta.$$

From the initial conditions we have $B_1 = \tfrac{3}{32}a_0^3$, $A_1 = a_1$; in the next step $a_1$ will be determined.

In the literature this process of applying the periodicity condition is sometimes called "elimination of the secular terms". The term "secular" originates from celestial mechanics and refers to terms which are unbounded with time; we have met with such terms in section 9.3.

Putting $a_0 = 2$ we find

$$\gamma_1(\theta) = a_1 \cos\theta + \frac{3}{4}\sin\theta - \frac{1}{4}\sin 3\theta.$$

The equation for $\gamma_2$ is

(10.10)
$$\gamma_2'' + \gamma_2 = (2\eta_1 + \frac{1}{4})\cos\theta + 2a_1 \sin\theta + 3a_1 \sin 3\theta +$$

$$-\frac{3}{2}\cos 3\theta + \frac{5}{4}\cos 5\theta).$$

The periodicity condition yields $(2\eta_1 + \frac{1}{4}) = 0, a_1 = 0$, which determines $\gamma_1(\theta)$ completely. The general solution of equation 10.10 is

$$\gamma_2(\theta) = A_2 \cos\theta + B_2 \sin\theta + \frac{3}{16}\cos 3\theta - \frac{5}{96}\cos 5\theta.$$

From the initial conditions we have $B_2 = 0$, $A_2 + \frac{13}{96} = a_2$; in the next step $a_2$ will be determined. The equation for $\gamma_3$ is

(10.11) $\qquad \gamma_3'' + \gamma_3 = 2\eta_2 \cos\theta + (2A_2 + \frac{1}{4})\sin\theta + (3A_2 - \frac{9}{32})\sin 3\theta$

$$+ \frac{35}{24}\sin 5\theta - \frac{7}{12}\sin 7\theta.$$

Note that the equations for $\gamma_n$ are all linear. This is in agreement with remark 2 in section 9.4.

The periodicity condition produces $\eta_2 = 0$, $2A_2 + \frac{1}{4} = 0$ which determines $a_2$. Using the relation $w^{-2} = 1 - \varepsilon\eta_0 - \varepsilon^2\eta_1 - \varepsilon^3\eta_2 - \cdots$ we find

$$w = 1 + \frac{1}{2}\varepsilon\eta_0 + \varepsilon^2(\frac{1}{2}\eta_1 + \frac{3}{8}\eta_0^2) + \varepsilon^3(\frac{1}{2}\eta_2 + \frac{3}{4}\eta_0\eta_1 + \frac{15}{48}\eta_0^3) + \cdots$$

For the period we have

$$T = \frac{2\pi}{w} = 2\pi(1 - \frac{1}{2}\varepsilon\eta_0 - \varepsilon^2(\frac{1}{2}\eta_1 + \frac{1}{8}\eta_0^2) - \varepsilon^3(\frac{1}{2}\eta_2 + \frac{1}{4}\eta_0\eta_1 +$$

$$\frac{3}{48}\eta_0^3) + \cdots).$$

For the van der Pol equation we find with $\eta_0 = 0$, $\eta_1 = -\frac{1}{8}, \eta_2 = 0$:

$$w = 1 - \frac{1}{16}\varepsilon^2 + O(\varepsilon^4)$$

$$T = 2\pi(1 + \frac{1}{16}\varepsilon^2) + O(\varepsilon^4).$$

The approximation of the periodic solution of the van der Pol equation to accuracy $O(\varepsilon^4)$

(10.12) $\qquad x(\theta) = 2\cos\theta + \varepsilon\{\frac{3}{4}\sin\theta - \frac{1}{4}\sin 3\theta\}+$

$$\varepsilon^2\{-\frac{1}{8}\cos\theta + \frac{3}{16}\cos 3\theta$$

$$-\frac{5}{96}\cos 5\theta\} + \varepsilon^3\{A_3 \cos\theta - \frac{7}{256}\sin\theta + \frac{21}{256}\sin 3\theta$$

$$-\frac{35}{576}\sin 5\theta + \frac{7}{576}\sin 7\theta\} + O(\varepsilon^4).$$

in which $\theta = (1 - \frac{1}{16}\varepsilon^2 t)$; the $O(\varepsilon^4)$ estimate is valid for $0 \le \theta \le 2\pi$ and $0 < \varepsilon \le \varepsilon_0$. We note that $\varepsilon_0$ is the radius of convergence of the power series with respect to $\varepsilon$ of the periodic solution. It is not so easy to compute this radius of convergence as we need for this the explicit form of the coefficients for arbitrary $n$ (or the ratio of subsequent coefficients).

In expression 10.12 there is a still unknown coefficient $A_3$, which can be computed when applying the periodicity condition to the solution of the equation for $\gamma_4$. Andersen and Geer (1982) used a computer program which formally manipulated the expansions; they calculated the expansion of $\omega$ to $O(\varepsilon^{164})$. Their result to $O(\varepsilon^{10})$ is

$$(10.13) \qquad \omega = 1 - \frac{1}{16}\varepsilon^2 + \frac{17}{3072}\varepsilon^4 + \frac{35}{884736}\varepsilon^6 - \frac{678899}{5096079360}\varepsilon^8 +$$

$$\frac{28160413}{2293235712000}\varepsilon^{10} + \cdots$$

The computations suggest a radius of convergence $\varepsilon_0 = \sqrt{3.42}$ which is rather large.

**Remark**

In example 7.4 we derived a criterion for the stability of the periodic solution of equations like van der Pol. In the case of the van der Pol equation we have stability if

$$\int_0^T (\phi^2(t) - 1)dt > 0.$$

The approximation given by 10.12 shows that the solution is stable.

Until now, the validity of the approximations obtained by expanding with respect to the small parameter $\varepsilon$ has been restricted to $0 \le \theta \le 2\pi$. This is not very long. In the next section we shall improve the validity of the approximations for periodic solutions strongly.

## 10.2  Approximation of periodic solutions on arbitrary long time-scales

When analysing a periodic solution of an equation with the Poincaré- Lindstedt method, the period and the other characteristic quantities (amplitude, phase) can be approximated with arbitrary good precision. One of the consequences is that we can find approximations which are valid on an interval of time which is much longer than the period. We shall illustrate the ideas again for equation 10.1 which has been discussed in the preceding section. In other problems where the Poincaré-Lindstedt method is being used, the reader can follow an analogous reasoning.

In the $x, \dot{x}$-phaseplane of equation 10.1, the $T$-periodic solution corresponds with a closed orbit; this also holds for the approximation to order $\varepsilon^N$ which we have constructed. During an interval of time of length $T$, the phase points corresponding with the periodic solution and the approximation are $\varepsilon^N$-close, we have almost synchronisation. After a longer period of time however, the phase points of the periodic solution and the approximation will separate more and more. We shall make this precise.

It follows from $\omega^{-2} = 1 - \varepsilon\eta(\varepsilon)$ that $\omega$ can be expanded in a power series, which is convergent for $0 < \varepsilon \leq \varepsilon_0$:

$$\omega = 1 + \sum_{n=1}^{\infty} \varepsilon^n \omega_n.$$

We now introduce the quantities

$$\theta_N = (1 + \varepsilon\omega_1 + \varepsilon^2\omega_2 + \ldots + \varepsilon^N\omega_N)t$$

and

$$x_M(\theta) = a_0 \cos\theta + \varepsilon\gamma_1(\theta) + \varepsilon^2\gamma_2(\theta) + \ldots + \varepsilon^M\gamma_M(\theta).$$

Note that in the Poincaré-Lindstedt method one has $M = N$ and the algorithm produces $x_N(\theta_N)$ which is an $O(\varepsilon^{N+1})$ approximation of the periodic solution on the time-scale 1. We shall demonstrate that the problem of "almost synchronisation" can be discussed on a longer time-scale by considering $M \neq N$. We shall use

$$x(\theta) - x_M(\theta_N) = x(\theta) - x(\theta_N) + x(\theta_N) - x_M(\theta_N)$$

and by application of the triangle inequality

(10.14)     $\|x(\theta) - x_M(\theta_N)\| \leq \|x(\theta) - x(\theta_N)\| + \|x(\theta_N) - x_M(\theta_N)\|.$

The function $x(\theta)$ is continuously differentiable, so, for $\varepsilon$ sufficiently small, $x(\theta)$ is Lipschitz-continuous in $\theta$ with Lipschitz-constant $L$. So we have

$$\|x(\theta) - x(\theta_N)\| \leq L\|\theta - \theta_N\| = L \left| \sum_{n=N+1}^{\infty} \varepsilon^n \omega_n \right| t.$$

Furthermore it follows from the expansion theorem that

$$x(\theta_N) - x_M(\theta_N) = \sum_{n=M+1}^{\infty} \varepsilon^n \gamma_n(\theta_N)$$

with bounded coefficients $\gamma_n$. Using these results, the estimate 10.14 becomes

(10.15)     $\|x(\theta) - x_M(\theta_N)\| \leq L \left| \sum_{n=N+1}^{\infty} \varepsilon^n \omega_n \right| t + \left| \sum_{n=M+1}^{\infty} \varepsilon^n \gamma_n(\theta_N) \right|$

$$= O(\varepsilon^{N+1}t) + O(\varepsilon^{M+1}).$$

For $M = N$ and $0 \leq t \leq T$ we recover the results of section 10.1 from 10.15. However with $M = N - 1$ we find for $0 \leq \varepsilon t \leq T$

$$\|x(\theta) - x_{N-1}(\theta_N)\| = O(\varepsilon^N) \text{ on the time-scale } 1/\varepsilon.$$

We also find

$$\|x(\theta) - x_{N-2}(\theta_N)\| = O(\varepsilon^{N-1}) \text{ on the time-scale } 1/\varepsilon^2.$$

The approximations and their corresponding accuracy can be summarised in table 10.1.

Table 10.1

| time-scale \ accuracy | $O(\varepsilon)$ | $O(\varepsilon^2)$ | $O(\varepsilon^3)$ |
|---|---|---|---|
| $0 \leq t \leq T$ | $a_0 \cos \theta_0$ | $a_0 \cos \theta_1 + \varepsilon \gamma_1(\theta_1)$ | $a_0 \cos \theta_2 + \varepsilon \gamma_1(\theta_2) + \varepsilon^2 \gamma_2(\theta_2)$ |
| $0 \leq \varepsilon t \leq T$ | $a_0 \cos \theta_1$ | $a_0 \cos \theta_2 + \varepsilon \gamma_1(\theta_2)$ | $a_0 \cos \theta_3 + \varepsilon \gamma_1(\theta_3) + \varepsilon^2 \gamma_2(\theta_3)$ |
| $0 \leq \varepsilon^2 t \leq T$ | $a_0 \cos \theta_2$ | $a_0 \cos \theta_3 + \varepsilon \gamma_1(\theta_3)$ | $a_0 \cos \theta_4 + \varepsilon \gamma_1(\theta_4) + \varepsilon^2 \gamma_2(\theta_4)$ |

In the table we have $\theta_0 = t$, $\theta_1 = (1+\varepsilon w_1)t$, $\theta_2 = (1+\varepsilon w_1+\varepsilon^2 w_2)t$, etc. The first row represents the convergent power series expansion which produces approximations, valid on the time-scale 1. The subsequent rows in the table contain asymptotic approximations which are valid on a longer time-scale.

We shall substitute the expressions in the table which apply to the periodic solution of the van der Pol equation $\ddot{x} + x = \varepsilon(1 - x^2)\dot{x}$:

Table 10.2. Periodic solution van der Pol equation

| time-scale \ accuracy | $O(\varepsilon)$ | $O(\varepsilon^2)$ | $O(\varepsilon^3)$ |
|---|---|---|---|
| $0 \leq t \leq T$ | $2 \cos t$ | $2 \cos t + \varepsilon \gamma_1(t)$ | $2 \cos \theta_2 + \varepsilon \gamma_1(\theta_2) + \varepsilon^2 \gamma_2(\theta_2)$ |
| $0 \leq \varepsilon t \leq T$ | $2 \cos t$ | $2 \cos \theta_2 + \varepsilon \gamma_1(\theta_2)$ | $2 \cos \theta_2 + \varepsilon \gamma_1(\theta_2) + \varepsilon^2 \gamma_2(\theta_2)$ |
| $0 \leq \varepsilon^2 t \leq T$ | $2 \cos \theta_2$ | $2 \cos \theta_2 + \varepsilon \gamma_1(\theta_2)$ | $2 \cos \theta_4 + \varepsilon \gamma_1(\theta_4) + \varepsilon^2 \gamma_2(\theta_4)$ |

in which

$$\theta_2 = (1 - \tfrac{1}{16}\varepsilon^2)t, \theta_4 = 1 - \tfrac{1}{16}\varepsilon^2 + \tfrac{17}{3072}\varepsilon^4$$
$$\gamma_1(\theta) = \tfrac{3}{4}\sin\theta - \tfrac{1}{4}\sin 3\theta$$
$$\gamma_2(\theta) = -\tfrac{1}{8}\cos\theta + \tfrac{3}{16}\cos 3\theta - \tfrac{5}{96}\cos 5\theta$$
$$T = 2\pi(1 + \tfrac{1}{16}\varepsilon^2 - \tfrac{5}{3072}\varepsilon^4).$$

Those approximations which are valid on the time-scale $1/\varepsilon$, can also be obtained by the averaging method which will be discussed in chapter 11.

Using the expansion, computed by Andersen and Geer (1982) which contains the expansion coefficients of $w$ to order $\varepsilon^{164}$, we can approximate the periodic solution on a very long time-scale. Take for instance $N = 10$. The function $2 \cos \theta_{10}$ with

$$\theta_{10} = (1 - \frac{1}{16}\varepsilon^2 + \frac{17}{3072}\varepsilon^4 + \frac{35}{884736}\varepsilon^6 - \frac{678899}{5096079360}\varepsilon^8 + \frac{28160413}{2293235712000}\varepsilon^{10})t$$

approximates the periodic solution of the van der Pol equation with accuracy $O(\varepsilon)$ on the time-scale $\varepsilon^{-10}$. If for instance $\varepsilon = 0.1$, the validity is on an interval of time of the order $10^{10}$ periods.

## 10.3 Periodic solutions of equations with forcing terms

The theory of nonlinear differential equations with inhomogeneous time-dependent terms, which represent oscillating systems with exciting forces, turns out to be very rich in phenomena. Here we consider only an important prototype problem, the forced Duffing-equation. This equation models many problems involving a nonlinear oscillator with damping and forcing.

The forced Duffing-equation will be considered in the form

$$(10.16) \qquad \ddot{x} + \varepsilon\mu\dot{x} + x - \varepsilon x^3 = \varepsilon h \cos \omega t$$

with constants $\mu \geq 0$ and $h > 0$; the forcing has a frequency which is near to 1 (a period which is near to $2\pi$) so we may put

$$\omega^{-2} = 1 - \varepsilon\beta$$

with $\beta$ a constant independent of $\varepsilon$. We shall look for solutions of equation 10.16 which are periodic with the period $2\pi/\omega$ of the forcing term.

Note that when studying autonomous equations, the period is not determined apriori (cf. section 10.1). Also, when studying autonomous equations, the translation property from lemma 2.1 holds; this enables us to put $\psi = 0$ for the phase $\psi$ in those cases without loss of generality. In the case of equation 10.16 we have not the translation property so it is reasonable to introduce a phase $\psi$. We put

$$\omega t = \theta - \psi$$

which, with $dx/d\theta = x'$, transforms equation 10.16 to

$$(10.17) \qquad \begin{aligned} x'' + x &= \varepsilon g(x, x', \psi, \theta, \varepsilon) \\ &= \varepsilon\beta x - \varepsilon\mu(1 - \varepsilon\beta)^{\frac{1}{2}}x' + \varepsilon(1 - \varepsilon\beta)x^3 + \\ &\quad + \varepsilon(1 - \varepsilon\beta)h \cos(\theta - \psi). \end{aligned}$$

We shall look for $2\pi$-periodic solutions of equation 10.17 with initial values $x(0) = a(\varepsilon)$, $x'(0) = 0$; this problem is aequivalent with looking for $2\pi$-periodic solutions of the integral equations

$$x(\theta) = a \cos \theta + \varepsilon \int_0^\theta \sin(\theta - \tau) g(x(\tau), x'(\tau), \psi, \tau, \varepsilon) d\tau$$

$$x'(\theta) = -a \sin \theta + \varepsilon \int_0^\theta \cos(\theta - \tau) g(x(\tau), x'(\tau), \psi, \tau, \varepsilon) d\tau.$$

The periodicity condition becomes

$$\int_0^{2\pi} \sin \tau g(x(\tau), x'(\tau), \psi, \tau, \varepsilon) d\tau = 0$$

(10.18)

$$\int_0^{2\pi} \cos \tau g(x(\tau), x'(\tau), \psi, \tau, \varepsilon) d\tau = 0.$$

Apart from $\varepsilon$, the periodic solution depends on $a$ and $\psi$. System 10.18 can be considered as a system of two equations with two unknowns, $a$ and $\psi$. According to the implicit function theorem, this system is uniquely solvable in a neighbourhood of $\varepsilon = 0$, if the corresponding determinant of Jacobi does not vanish. Rewriting system 10.18 as $F_1(a, \psi) = 0$, $F_2(a, \psi) = 0$ this means that we have unique solvability if

(10.19)
$$\frac{\partial(F_1, F_2)}{\partial(a, \psi)} \neq 0,$$

with $\varepsilon$ in a neighbourhood of 0.

According to the implicit function theorem we may expand

$$a(\varepsilon) = \sum_{n=0}^{\infty} \varepsilon^n a_n, \psi(\varepsilon) = \sum_{n=0}^{\infty} \varepsilon^n \psi_n, x(\theta) = a_0 \cos \theta + \sum_{n=1}^{\infty} \varepsilon^n \gamma_n(\theta).$$

Using these series, the expression for $g$ in 10.17 yields the expansion

$$g(x, x', \psi, \theta, \varepsilon) = \beta a_0 \cos \theta + \mu a_0 \sin \theta + a_0^3 \cos^3 \theta + h \cos(\theta - \psi_0) + \varepsilon \dots$$

The periodicity condition 10.18 becomes at first order

(10.20)
$$\mu a_0 + h \sin \psi_0 = 0$$
$$\beta a_0 + h \cos \psi_0 + \tfrac{3}{4} a_0^3 = 0.$$

System 10.20 is uniquely solvable if (10.19)

(10.21)
$$\mu \sin \psi_0 + \beta \cos \psi_0 + \frac{9}{4} a_0^2 \cos \psi_0 \neq 0.$$

The problem without damping, $\mu = 0$, is the simplest. We have in this case $\psi_0 = 0, \pi$; the possible values of $a_0$ follow from the equations $\beta a_0 \pm h + \tfrac{3}{4} a_0^3 = 0$. The requirement of uniqueness 10.21 becomes in this case $\beta + \tfrac{9}{4} a_0^2 \neq 0$.
If, for given values of $\mu, h$ and $\beta$, we have determined the coefficients $a_0$ and $\psi_0$, we find for the periodic solution

$$x(t) = a_0 \cos(\omega t + \psi_0) + O(\varepsilon) \text{ on the time-scale 1.}$$

From the discussion in the preceding section it is clear that, if we wish to construct an approximation which is valid on a longer time-scale, we have to compute more terms of the expansion.

Another remark is, that we have obtained an approximation of a periodic solution but until now there is no simple criterion to determine its stability at the same time. We shall return to this question in the next chapter.

## 10.4    The existence of periodic solutions

In section 10.1 we studied expansion with respect to the small parameter $\varepsilon$ for periodic solutions of the equation $\ddot{x} + x = \varepsilon f(x, \dot{x}, \varepsilon)$; in section 10.2 we considered a problem where the righthand side is time-dependent.

In both cases we were able to apply the Poincaré expansion theorem of section 9.4. The periodicity condition has the form $F_1(a, \eta) = 0$, $F_2(a, \eta) = 0$ (equations 10.3 or 10.18) with $a$ and $\eta$ two parameters, which still have to be determined. The equations for $a$ and $\eta$ are uniquely solvable if

(10.22)
$$\frac{\partial(F_1, F_2)}{\partial(a, \eta)} \neq 0.$$

This is the most important condition for application of the implicit function theorem, the other conditions of this theorem have been satisfied by the assumptions of the Poincaré expansion theorem.

We note that in principle, condition 10.22 is difficult to verify as the unknown periodic solution is contained in the expressions for $F_1$ and $F_2$. However, verification of condition 10.22 is possible by applying the expansion theorem to develop $F_1$ and $F_2$ with respect to the small parameter. We summarise this as follows.

**Theorem 10.1**

Consider equation 10.1 $\ddot{x} + x = \varepsilon f(x, \dot{x}, \varepsilon)$ (or equation 10.16). If the conditions of the Poincaré expansion theorem have been satisfied and simultaneously the periodicity condition 10.3 (or 10.18) and the uniqueness condition 10.22, then there exists a periodic solution. This solution can be represented by a convergent power series in $\varepsilon$ for $0 \leq \varepsilon < \varepsilon_0$.

The problem of the forced Duffing equation can be generalised as follows. Consider in $\mathbf{R}^n$ the equation

(10.23)
$$\dot{x} = f(t, x) + \varepsilon g(t, x, \varepsilon).$$

with (for instance) $f$ and $g$ $T$-periodic. The "unperturbed equation" is

(10.24)
$$\dot{y} = f(t, y)$$

and this equation is assumed to have a $T$-periodic solution $\phi(t)$. Does there exist a $T$-periodic solution $\psi(t, \varepsilon)$ which is close to $\phi(t)$ with $\psi(t, 0) = \phi(t)$? Consider equation 10.23 with initial condition

(10.25)
$$x \mid_{t=0} = \phi(0) + \mu.$$

The solution of this initial value problem will be $x(t, \varepsilon, \mu)$. We assume that the conditions of the expansion theorem have been satisfied so that

(10.26)
$$x(t, \varepsilon, \mu) = \phi(t) + \sum_{k_1 + \ldots k_n + k \geq 1} \gamma_{k_1 \ldots k_n k(t)} \mu_1^{k_1} \ldots \mu_n^{k_n} \varepsilon^k.$$

The series converges for $\|\mu\| \le \rho$, $0 \le \varepsilon < \varepsilon_0$ and $t_0 \in [0,T]$; $\mu_1, \ldots, \mu_n$ are the components of $\mu$. The coefficients $\gamma$ can be determined by substitution of the series 10.26 in equation 10.23 and by equating coefficients of equal powers of the parameters on both sides of the equation. This process produces a system of linear differential equations for the unknown expansion coefficients $\gamma$ which presumably we can solve.

The periodicity condition is

(10.27) $$H(\mu_1, \ldots, \mu_n, \varepsilon) = x(T, \varepsilon, \mu) - x(0, \varepsilon, \mu) = 0.$$

This system of $n$ equations with $n$ unknowns $\mu_1, \ldots, \mu_n$ is uniquely solvable for $\varepsilon$ near zero if for the Jacobian we have

$$\frac{\partial H}{\partial \mu} \ne 0.$$

The corresponding solutions for the parameters $\mu_1, \ldots, \mu_n$ of 10.27 produce a $T$-periodic solution 10.26 of equation 10.23.

In applying the Poincaré-Lindstedt method all kinds of complications may arise. For instance in a number of problems, like in Hamiltonian systems, periodic solutions arise in continuous one-parameter families instead of isolated periodic solutions. In such a case condition 10.22 will not be satisfied and one has to take a closer look at the question whether the expansions which have been obtained represent periodic solutions.

The verification of condition 10.22 can usually be carried out only by expansion with respect to the small parameter. In the example of section 10.1, condition 10.22 has been satisfied already at first order in the expansion. In practice it may happen that one needs higher order expansions to achieve this. We illustrate this as follows.

**Example 10.2**

Consider the modified van der Pol equation for $\varepsilon > 0$

(10.28) $$\ddot{x} + x = \varepsilon x^3 + \varepsilon^2 (1 - x^2)\dot{x}.$$

We shall look for periodic solutions of this equation, but the theory of chapter 4 doesn't present us with apriori knowledge of the existence of such a solution. As in example 10.1 we put

$$\omega t = \theta, \omega^{-2} = 1 - \varepsilon \eta(\varepsilon).$$

Equation 10.28 becomes with $dx/d\theta = x'$

(10.29) $$x'' + x = \varepsilon[\eta x + (1 - \varepsilon\eta)x^3 + \varepsilon(1 - \varepsilon\eta)^{\frac{1}{2}}(1 - x^2)x'].$$

The initial values will be again $x(0) = a(\varepsilon), x'(0) = 0$. We expand

$$a(\varepsilon) = \sum_{n=0}^{\infty} \varepsilon^n a_n, \quad \eta(\varepsilon) = \sum_{n=0}^{\infty} \varepsilon^n \eta_n$$

$$\begin{aligned}
x(\theta) &= a_0 \cos \theta + \varepsilon \gamma_1(\theta) + \varepsilon^2 \gamma_2(\theta) + \ldots \\
x'(\theta) &= -a_0 \sin \theta + \varepsilon \gamma_1'(\theta) + \varepsilon^2 \gamma_2'(\theta) + \ldots
\end{aligned}$$

Substitution in equation 10.29 and equating coefficients of equal powers of $\varepsilon$ yields

(10.30) $\qquad \gamma_1'' + \gamma_1 = \eta_0 a_0 \cos \theta + a_0^3 \cos^3 \theta$

(10.31) $\qquad \gamma_2'' + \gamma_2 = \eta_1 a_0 \cos \theta + \eta_0 \gamma_1(\theta) - \eta_0 a_0^3 \cos^3 \theta$

$$+ 3a_0^2 \cos^2 \theta \gamma_1(\theta) - (1 - a_0^2 \cos^2 \theta) a_0 \sin \theta$$

etc.

The general solution of equation 10.30 is

$$\gamma_1(\theta) = A_1 \cos \theta + B_1 \sin \theta + \int_0^\theta \sin(\theta - \tau)[\eta_0 a_0 \cos \tau +$$

$$+ a_0^3 \cos^3 \tau] d\tau.$$

From the initial values we have $A_1 = a_1$, $B_1 = 0$; $a_0$ and $a_1$ are as yet unknown. The periodicity condition produces

$$\int_0^{2\pi} \sin \tau [\eta_0 a_0 \cos \tau + a_0^3 \cos^3 \tau] d\tau = 0$$

(10.32)

$$\int_0^{2\pi} \cos \tau [\eta_0 a_0 \cos \tau + a_0^3 \cos^3 \tau] d\tau = 0$$

The first equation of 10.32 holds for all values of $\eta_0$ and $a_0$; from the second equation we find

$$a_0 \left( \eta_0 + \frac{3}{4} a_0^2 \right) = 0.$$

The Jacobian 10.22 of system 10.32 is zero. The parameters $a_0$ and $\eta_0$ have not been determined uniquely. We find with $\eta_0 = -\frac{3}{4} a_0^2$

$$\gamma_1(\theta) = a_1 \cos \theta + \int_0^\theta \sin(\theta - \tau) \frac{a_0^3}{4} \cos 3\tau \, d\tau$$

$$= a_1 \cos \theta + \frac{a_0^3}{32} (\cos \theta - \cos 3\theta).$$

The reader should consider the consequences of the choice $a_0 = 0$.

In the same way we determine the solution of equation 10.31. The periodicity condition produces

$$\int_0^{2\pi} \sin\tau [\eta_1 a_0 \cos\tau + \eta_0 \gamma_1(\tau) - \eta_0 a_0^3 \cos^3\tau +$$

$$3a_0^2 \cos^2\tau \gamma_1(\tau) - (1 - a_0^2 \cos^2\tau) a_0 \sin\tau] d\tau = 0$$

(10.33)

$$\int_0^{2\pi} \cos\tau [\eta_1 a_0 \cos\tau + \eta_0 \gamma_1(\tau) - \eta_0 a_0^3 \cos^3\tau +$$

$$3a_0^2 \cos^2\tau \gamma_1(\tau) - (1 - a_0^2 \cos^2\tau) a_0 \sin\tau] d\tau = 0.$$

Condition 10.33 leads to

$$a_0 \left(1 - \frac{1}{4} a_0^2\right) = 0$$

$$\eta_1 a_0 + \eta_0 a_1 - \frac{23}{32} \eta_0 a_0^3 + \frac{9}{4} a_0^2 a_1 + \frac{a_0^3}{64} = 0$$

It follows that $a_0 = 2$ so that $\eta_0 = -3$. In this way the first term of the approximation of $x(\theta)$ has been determined uniquely. The second relation gives

$$2\eta_1 + 6a_1 + \frac{139}{8} = 0.$$

The parameters $\eta_1$ and $a_1$ can be determined by analysing the periodicity condition for the solution of the equation for $\gamma_2(\theta)$. The periodic solution of equation 10.28 is approximated as

$$x(t) = 2\cos(1 - \frac{3}{2}\varepsilon)t + O(\varepsilon) \text{ on the time-scale } 1/\varepsilon.$$

## 10.5  Exercises

10-1. Consider the equation

$$\ddot{x} - x - x^2 = 0$$

  a. Characterise the critical points and sketch the phase-plane.

  b. Around the critical point $(-1, 0)$ we have cycles which correspond with periodic solutions.
  Starting in a neighbourhood at $(x, \dot{x}) = (-1 + \varepsilon, 0)$ we expand the period

$$T(\varepsilon) = T_0 + \varepsilon T_1 + \varepsilon^2 T_2 + \varepsilon^3 \dots$$

  Determine $T_0, T_1$ and $T_2$.

10-2. The solutions of the equation for the mathematical pendulum $\ddot{x} + \sin x = 0$ are periodic in a neighbourhood of $(x, \dot{x}) = (0, 0)$. Figure 5.2 suggests that the period $T$ of the solutions starting in $(a, 0)$ increases slowly with $a$ (if this is true, it explains the success of the harmonic oscillator as an approximation of the mathematical pendulum).

To verify this, consider $a$ as a small parameter and approximate $T(a)$.

10-3. We left out some details of the calculation of appendix 2, case $m = 2$. Find the intermediate steps which have been omitted.

10-4. Consider the van der Pol-equation with a small forcing

$$\ddot{x} + x = \varepsilon(1 - x^2)\dot{x} + \varepsilon h \cos(wt)$$

We shall look for periodic solutions with initial conditions $x(0) = a(\varepsilon)$, $\dot{x}(0) = 0$. Determine the Poincaré-Lindstedt expansion to $O(\varepsilon)$ in the case $w^{-2} = 1 - \varepsilon\beta$ by computing the periodicity conditions.

# 11  The method of averaging

## 11.1  Introduction

In this chapter we shall consider again equations containing a small parameter $\varepsilon$. The approximation method leads generally to asymptotic series as opposed to the convergent series studied in the preceding chapter; see section 9.2 for the basic concepts. This asymptotic character of the approximations is more natural in many problems; also the method turns out to be very powerful, it is not restricted to periodic solutions.

The idea of averaging as a computational technique, without proof of validity, originates from the $18^{th}$ century; it has been formulated very clearly by Lagrange (1788) in his study of the gravitational three-body problem as a perturbation of the two-body problem.

As an example we consider the equation

$$(11.1) \qquad\qquad \ddot{x} + x = \varepsilon f(x, \dot{x}).$$

If $\varepsilon = 0$, the solutions are known: a linear combination of $\cos t$ and $\sin t$. We can also write this linear combination as

$$r_0 \cos(t + \psi_0).$$

The amplitude $r_0$ and the phase $\psi_0$ are constants and they are determined by the initial values. To study the behaviour of the solutions for $\varepsilon \neq 0$, Lagrange introduces "variation of constants". One assumes that for $\varepsilon \neq 0$, the solution can still be written in this form where the amplitude $r$ and the phase $\psi$ are now functions of time. So we put for the solution of equation 11.1

$$(11.2) \qquad\qquad x(t) = r(t) \cos(t + \psi(t))$$

$$(11.3) \qquad\qquad \dot{x}(t) = -r(t) \sin(t + \psi(t)).$$

Substitution of these expressions for $x$ and $\dot{x}$ into equation 11.1 produces an equation for $r$ and $\psi$

$$-\dot{r} \sin(t + \psi) - r \cos(t + \psi)(1 + \dot{\psi}) + r \cos(t + \psi) =$$
$$= \varepsilon f(r \cos(t + \psi), -r \sin(t + \psi))$$

or

$$(11.4) \qquad -\dot{r} \sin(t + \psi) - r \cos(t + \psi)\dot{\psi} = \varepsilon f(r \cos(t + \psi), -r \sin(t + \psi)).$$

Another requirement is that differentiation of the righthandside of equation 11.2 must produce an expression which equals the righthand side of equation 11.3. So we find

$$\dot{r}\cos(t+\psi) - r\sin(t+\psi)(1+\dot{\psi}) = -r\sin(t+\psi)$$

or

(11.5) $$\dot{r}\cos(t+\psi) - \dot{\psi}r\sin(t+\psi) = 0.$$

Equations 11.4 and 11.5 can be considered as two algebraic equations in $\dot{r}$ and $\dot{\psi}$; the solutions are

$$\dot{r} = -\varepsilon\sin(t+\psi)f(r\cos(t+\psi), -r\sin(t+\psi))$$

(11.6)

$$\dot{\psi} = -\frac{\varepsilon}{r}\cos(t+\psi)f(r\cos(t+\psi), -r\sin(t+\psi)).$$

The transformation 11.2-3 $x, \dot{x} \to r, \psi$ presupposes of course that $r \neq 0$; in problems where $r(t_0) = 0$ we have to use a different transformation.

The system of differential equations 11.6 is supplied with initial conditions, for instance at $t_0 = 0$ determined by

$$x(0) = r(0)\cos\psi(0)$$

$$\dot{x}(0) = -r(0)\sin\psi(0).$$

The reasoning of Lagrange is now as follows. The righthand sides of system 11.6 can be expanded in the form

$$\varepsilon[g_0(r,\psi) + g_1(r,\psi)\sin t + h_1(r,\psi)\cos t + \ldots]$$

where the dots represent the higher order harmonics $g_n(r,\psi)\sin nt$ and $h_n(r,\psi)\cos nt$, $n = 2, 3, \ldots$ So this is what we are calling nowadays a Fourier expansion of the righthand sides. According to equations 11.6 the variables $r$ and $\psi$ change slowly with time; the contribution of these changes is, averaged over the period $2\pi$, zero, except for the term $g_0(r,\psi)$. So we omit all the terms which have average zero and we solve the simplified system of differential equations.

The idea of averaging is so natural, that for a long time the method has been used in many fields of application without people bothering about proofs of validity. Until the end of the 19th century the main field of application was celestial mechanics; around 1920 van der Pol promoted the use of the method for equations arising in electronic circuit theory. Only in 1928, the first proof of asymptotic validity of the method was given by Fatou. After 1930 Krylov, Bogoliubov and Mitropolsky of the Kiev school of mathematics pursued this type of research, see Bogoliubov and Mitropolsky (1961).

A survey of the theory of averaging and many new results can be found in Sanders and Verhulst (1985).
We conclude this introduction with an explicit calculation.

**Example 11.1**
Consider the equation for a nonlinear oscillator with damping

(11.7) $$\ddot{x} + x = \varepsilon(-\dot{x} + x^2).$$

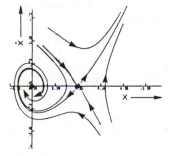

*Figure 11.1*

The trivial solution is asymptotically stable, see example 8.6. We shall apply the method of Lagrange. Transformation 11.2-3 yields

$$\begin{aligned} \dot{r} &= -\varepsilon r \sin^2(t+\psi) - \varepsilon r^2 \sin(t+\psi)\cos^2(t+\psi) \\ \dot{\psi} &= -\varepsilon \cos(t+\psi)\sin(t+\psi) - \varepsilon r \cos^3(t+\psi). \end{aligned}$$

The righthand sides are $2\pi$-periodic in $t$. We average now over $t$, keeping $r$ and $\psi$ fixed. The result is

$$\dot{r}_a = -\frac{1}{2}\varepsilon r_a , \quad \dot{\psi}_a = 0$$

with index $a$ to indicate that we approximate $r$ and $\psi$ by $r_a$ and $\psi_a$. The averaged equations are very easy to solve:

$$r_a(t) = r(0)e^{-\frac{1}{2}\varepsilon t} , \quad \psi_a(t) = \psi(0).$$

As an approximation of $x(t)$ we propose

(11.8) $$x_a(t) = r(0)e^{-\frac{1}{2}\varepsilon t}\cos(t+\psi(0)).$$

For which initial values does expression 11.8 appear to be a reasonable approximation of the solution (see figure 11.1 for the phase-plane of equation 11.7)?

## 11.2 The Lagrange standard form

In proving the asymptotic validity of the approximations one usually starts off with a slowly varying system like equation 11.6. Here we show how to obtain such equations by the method of "variation of parameters".
Consider for $x \in R^n$, $t \geq 0$ the initial value problem

(11.9) $$\dot{x} = A(t)x + \varepsilon g(t,x), \quad x(0) = x_0.$$

$A(t)$ is a continuous $n \times n$-matrix, $g(t,x)$ is a sufficiently smooth function of $t$ and $x$. The unperturbed ($\varepsilon = 0$) equation is linear and it has $n$ independent solutions which are used to compose a fundamental matrix $\phi(t)$. We put

$$x = \Phi(t)y \quad \text{(Lagrange)}$$

In mechanics this is called sometimes "introducing comoving coordinates".
Substitution in equation 11.9 produces

$$\dot{\Phi}y + \Phi\dot{y} = A(t)\Phi y + \varepsilon g(t, \Phi y)$$

and as $\dot{\Phi} = A(t)\Phi$ we have

$$\Phi\dot{y} = \varepsilon g(t, \Phi y).$$

Equations 11.4-5 are a special case of this vector equation. We solve for $\dot{y}$ by inverting the fundamental matrix :

(11.10) $$\dot{y} = \varepsilon \Phi^{-1}(t)g(t, \Phi(t)y).$$

The initial values follow from

$$y(0) = \Phi^{-1}(0)x_0.$$

Writing down the equations for the components of $y$ while using Cramer's rule we have

(11.11) $$\dot{y}_i = \varepsilon \frac{W_i(t,y)}{W(t)}, \quad i = 1, \ldots, n,$$

with $W(t) = |\ \Phi(t)\ |$, the wronskian (determinant of $\Phi(t)$); $W_i(t,y)$ is the determinant of the matrix which one obtains by replacing the $i^{th}$ column of $\Phi(t)$ by $g$. Equation 11.10 has the so-called (Lagrange) standard form. More in general the standard form is

$$\dot{y} = \varepsilon f(t,y).$$

**Remark**
The unperturbed equation in the case of equation 11.9 is linear; this facilitates the variation of parameters procedure. If the unperturbed equation is nonlinear the variation of parameters technique still applies. In practice however there are usually many technical obstructions while carrying out the procedure.

## Example 11.2

Consider the equation

(11.12) $$\ddot{x} + \omega^2 x = \varepsilon g(t, x, \dot{x})$$

with constant $\omega > 0$. We have some freedom in choosing the transformation to the standard form as there are several representations of the solutions of the unperturbed problem. We start with an amplitude-phase transformation

(11.13) $$\begin{aligned} x(t) &= r(t)\cos(\omega t + \psi(t)) \\ \dot{x}(t) &= -r(t)\omega \sin(\omega t + \psi(t)) \end{aligned}$$

We find the equations

(11.14) $$\begin{aligned} \dot{r} &= -\frac{\varepsilon}{\omega}\sin(\omega t + \psi)g(t, r\cos(\omega t + \psi), -r\omega\sin(\omega t + \psi)) \\ \dot{\psi} &= -\frac{\varepsilon}{\omega r}\cos(\omega t + \psi)g(t, r\cos(\omega t + \psi), -r\omega\sin(\omega t + \psi)). \end{aligned}$$

Another possibility is to use the transformation

(11.15) $$\begin{aligned} x(t) &= y_1(t)\cos\omega t + \frac{y_2(t)}{\omega}\sin\omega t \\ \dot{x}(t) &= -\omega y_1(t)\sin\omega t + y_2(t)\cos\omega t. \end{aligned}$$

We find the equations

(11.16) $$\begin{aligned} \dot{y}_1 &= -\frac{\varepsilon}{\omega}\sin\omega t\, g(t, x(t), \dot{x}(t)) \\ \dot{y}_2 &= \varepsilon\cos\omega t\, g(t, x(t), \dot{x}(t)). \end{aligned}$$

In $g$ the expressions for $x(t)$ and $\dot{x}(t)$ still have to be substituted.
Equation 11.12 can be treated both with the standard form 11.14 and the standard form 11.16. It turns out that the explicit form of the righthand side $\varepsilon g(t, x, \dot{x})$ determines which choice is the best one.

## 11.3  Averaging in the periodic case

We shall now consider the asymptotic validity of the averaging method. Consider the initial value problem

(11.17) $$\dot{x} = \varepsilon f(t, x) + \varepsilon^2 g(t, x, \varepsilon), \quad x(0) = x_0.$$

We assume that $f(t, x)$ is $T$-periodic in $t$ and we introduce the average

$$f^0(y) = \frac{1}{T}\int_0^T f(t, y)dt.$$

In performing the integration $y$ has been kept constant. Consider now the initial value problem for the averaged equation

(11.18) $$\dot{y} = \varepsilon f^0(y), \quad y(0) = x_0.$$

The vector function $y(t)$ represents an approximation of $x(t)$ in the following way :

**Theorem 11.1**

Consider the initial value problems 11.17 and 11.18 with $x, y, x_0 \in D \subset \mathbf{R}^n, t \geq 0$.
Suppose that

   a. the vector functions $f, g$ and $\partial f / \partial x$ are defined, continuous and bounded by a constant $M$ (independent of $\varepsilon$) in $[0, \infty) \times D$;

   b. $g$ is Lipschitz-continuous in $x$ for $x \in D$;

   c. $f(t, x)$ is $T$-periodic in $t$ with average $f^0(x)$; $T$ is a constant which is independent of $\varepsilon$;

   d. $y(t)$ is contained in an internal subset of $D$.

Then we have $x(t) - y(t) = O(\varepsilon)$ on the time-scale $1/\varepsilon$.

**Proof**

The assumptions $a$ and $b$ guarantee the existence and uniqueness of the solutions of problems 11.17 and 11.18 on the time-scale $1/\varepsilon$ (see theorem 1.1).
We introduce

(11.19)
$$u(t, x) = \int_0^t [f(s, x) - f^0(x)] ds.$$

As we subtract the average of $f(s, x)$ in the integrand, the integral is bounded :

$$\|u(t, x)\| \leq 2MT , \ t \geq 0, x \in D.$$

We now introduce a "near-identity transformation"

(11.20)
$$x(t) = z(t) + \varepsilon u(t, z(t)).$$

We call this "near-identity" as $x(t) - z(t) = O(\varepsilon)$ for $t \geq 0$, $x, z \in D$. Transformation 11.20 will be used to simplify equation 11.17; this is also called *normalisation*, see sections 13.2 and 13.3.
Differentiation of 11.20 and substitution in 11.17 yields

$$\dot{x} = \dot{z} + \varepsilon \frac{\partial}{\partial t} u(t, z) + \varepsilon \frac{\partial}{\partial z} u(t, z)\dot{z} = \varepsilon f(t, z + \varepsilon u(t, z)) +$$

$$+ \varepsilon^2 g(t, z + \varepsilon u(t, z), \varepsilon).$$

Using 11.19 we write this equation in the form

$$[I + \varepsilon \frac{\partial}{\partial z} u(t, z)]\dot{z} = \varepsilon f^0(z) + R$$

with $I$ the $n \times n$-identity matrix; $R$ is short for the vector function

$$R = \varepsilon f(t, z + \varepsilon u(t, z)) - \varepsilon f(t, z) + \varepsilon^2 g(t, z + \varepsilon u(t, z), \varepsilon).$$

$\partial u / \partial z$ is uniformly bounded (as $u$) so we may invert to obtain

$$(11.21) \qquad [I + \varepsilon \frac{\partial}{\partial z} u(t, z)]^{-1} = I - \varepsilon \frac{\partial}{\partial z} u(t, z) + O(\varepsilon^2), \ t \geq 0, z \in D.$$

From the Lipschitz-continuity of $f(t, z)$ we have

$$\begin{aligned} \|f(t, z + \varepsilon u(t, z)) - f(t, z)\| &\leq L\varepsilon \|u(t, z)\| \\ &\leq L\varepsilon 2MT. \end{aligned}$$

$L$ is the Lipschitz-constant. Because of the boundedness of $g$ it follows that for some positive constant $C$, independent of $\varepsilon$, we have the estimate

$$(11.22) \qquad \|R\| \leq \varepsilon^2 C , \ t \geq 0, \ z \in D.$$

With 11.21 and 11.22 we find for $z$

$$(11.23) \qquad \dot{z} = \varepsilon f^0(z) + R - \varepsilon^2 \frac{\partial u}{\partial z} f^0(z) + O(\varepsilon^3) , \ z(0) = x(0).$$

As $R = O(\varepsilon^2)$ we may put equation 11.23 in the form corresponding with theorem 9.1 by introducing the time-like variable $\tau = \varepsilon t$. We conclude that the solution of

$$\frac{dy}{d\tau} = f^0(y) , \ y(0) = z(0)$$

approximates the solution of equation 11.23 with error $O(\varepsilon)$ on the time-scale 1 in $\tau$, i.e. on the time-scale $1/\varepsilon$ in $t$. Because of the near-identity transformation 11.20 the same estimate holds for $y(t)$ as an approximation of $x(t)$. □

## Remark 1

In example 11.1 we considered the equation

$$\ddot{x} + x = -\varepsilon \dot{x} + \varepsilon x^2$$

and we obtained by averaging the expression

$$x_a(t) = r(0) e^{-\frac{1}{2}\varepsilon t} \cos(t + \psi(0)).$$

From the theorem which we proved just now, we conclude that

$$x(t) - x_a(t) = O(\varepsilon) \text{ on the time-scale } 1/\varepsilon.$$

The initial values $r(0)$ and $\psi(0)$ are supposed to be $O(1)$ quantities with respect to $\varepsilon$. So the estimates are not valid if we start near the saddle point $x = 1/\varepsilon , \dot{x} = 0$.

**Remark 2**

It is possible to carry out averaging in the periodic case and to prove asymptotic validity under weaker assumptions, for instance in the case that the vector function $f$ is not differentiable. On the other hand, on adding assumptions with respect to the vector functions $f$ and $g$, we can construct second and higher order approximations. A second order approximation is generally characterised by an error $O(\varepsilon^2)$ on the time-scale $1/\varepsilon$.

**Example 11.3**

Consider the autonomous equation

(11.24)
$$\ddot{x} + x = \varepsilon f(x, \dot{x})$$

with given initial values. Transforming $x, \dot{x} \rightarrow r, \psi$ with 11.2-3 we obtain the equations

$$\dot{r} = -\varepsilon \sin(t + \psi) f(r \cos(t + \psi), -r \sin(t + \psi))$$

$$\dot{\psi} = -\frac{\varepsilon}{r} \cos(t + \psi) f(r \cos(t + \psi), -r \sin(t + \psi)).$$

The righthand side is $2\pi$-periodic in $t$; while averaging $\psi$ is kept constant, so for the averaging process the system can also be considered to be $2\pi$-periodic in $t + \psi$. With $s = t + \psi$ we find

$$\frac{1}{2\pi} \int_0^{2\pi} \sin(t + \psi) f(r \cos(t + \psi), -r \sin(t + \psi)) dt =$$

$$\frac{1}{2\pi} \int_\psi^{2\pi + \psi} \sin s f(r \cos s, -r \sin s) ds = f_1(r).$$

Because of the $2\pi$-periodicity of the integrand, $f_1(r)$ doesnot depend on $\psi$. In the same way we find

$$\frac{1}{2\pi} \int_0^{2\pi} \cos(t + \psi) f(r \cos(t + \psi), -r \sin(t + \psi)) dt = f_2(r).$$

An asymptotic approximation $r_a, \psi_a$ of $r, \psi$ can be found as solutions of

(11.25)
$$\begin{aligned} \dot{r}_a &= -\varepsilon f_1(r_a), \ r_a(0) = r(0) \\ \dot{\psi}_a &= -\frac{\varepsilon}{r_a} f_2(r_a), \ \psi_a(0) = \psi(0). \end{aligned}$$

As $\psi_a$ is not present in the righthand sides, the order of the differential equations to be solved has been reduced to one. We apply this to a standard example, the van der Pol-equation

$$\ddot{x} + x = \varepsilon(1 - x^2)\dot{x}.$$

Using the formulas in appendix 3 we find

(11.26) $$\dot{r}_a = \frac{1}{2}\varepsilon r_a\left(1 - \frac{1}{4}r_a^2\right), \quad \dot{\psi}_a = 0.$$

Taking $r(0) = 2$ we have $r_a(t) = 2$, $t \geq 0$ (critical point of the first equation in 11.26). This initial value corresponds with the periodic solution which we studied earlier with the Poincaré-Lindstedt method in example 10.1. Using averaging we find with $r(0) = 2$, $\psi(0) = 0$

$$x(t) = 2\cos t + O(\varepsilon) \text{ on the time-scale } 1/\varepsilon.$$

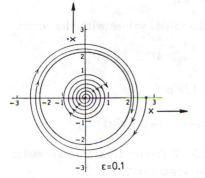

*Figure 11.2*

By the averaging method we find moreover an approximation for the other solutions by solving equations 11.26:

$$r_a(t) = \frac{r(0)e^{\frac{1}{2}\varepsilon t}}{[1 + \frac{1}{4}r_0^2(e^{\varepsilon t} - 1)]^{\frac{1}{2}}}, \quad \psi_a(t) = \psi(0).$$

As $x_a(t) = r_a(t)\cos(t + \psi_a(t))$ we find that with increasing time $t$, the solutions approach the periodic solution. In figure 11.2 we have presented the phase-plane in the case $\varepsilon = .1$. Compare this phase-plane with case $\varepsilon = 1$ in figure 2.9.

**Example 11.4**
Consider the Mathieu-equation

(11.27) $$\ddot{x} + (1 + 2\varepsilon\cos 2t)x = 0$$

with initial values $x(0) = x_0$, $\dot{x}(0) = 0$. Equation 11.27 models linear oscillators in engineering which are subjected to frequency modulation. We apply the transformations from example 11.2 with $g = -2\varepsilon\cos(2t)x$ and $\omega = 1$. Introducing the amplitude $r$ and phase $\psi$, the averaged equations become rather complicated. In equations like this it is more convenient to use transformation 11.15

$$\begin{aligned} x(t) &= y_1(t)\cos t + y_2(t)\sin t \\ \dot{x}(t) &= -y_1(t)\sin t + y_2(t)\cos t \end{aligned}$$

which leads to

$$\begin{aligned} \dot{y}_1 &= 2\varepsilon\sin t\cos 2t(y_1\cos t + y_2\sin t), \quad y_1(0) = x_0 \\ \dot{y}_2 &= -2\varepsilon\cos t\cos 2t(y_1\cos t + y_2\sin t), \quad y_2(0) = 0. \end{aligned}$$

The righthand side is $2\pi$-periodic in $t$. We find for the averaged equations (appendix 3) :

$$\dot{y}_{1a} = -\frac{1}{2}\varepsilon y_{2a}\,, \ \dot{y}_{2a} = -\frac{1}{2}\varepsilon y_{1a}.$$

We solve these linear equations while imposing the initial values with the result

$$y_{1a}(t) = \frac{1}{2}x_0 e^{-\frac{1}{2}\varepsilon t} + \frac{1}{2}x_0 e^{\frac{1}{2}\varepsilon t}, y_{2a}(t) = \frac{1}{2}x_0 e^{-\frac{1}{2}\varepsilon t} - \frac{1}{2}x_0 e^{\frac{1}{2}\varepsilon t}.$$

An $O(\varepsilon)$ approximation of $x(t)$ on the time-scale $1/\varepsilon$ is

$$x_a(t) = \frac{1}{2}x_0 e^{-\frac{1}{2}\varepsilon t}(\cos t + \sin t) + \frac{1}{2}x_0 e^{\frac{1}{2}\varepsilon t}(\cos t - \sin t).$$

If $\varepsilon = 0$, the solutions of equation 11.27 are all stable. It follows from the expression for $x_a(t)$ that this is not the case if $\varepsilon > 0$; see also appendix 2.

## 11.4   Averaging in the general case

We shall now consider the case in which the vector function $f(t,x)$ is not periodic but where the average exists in a more general sense. A simple example which arises often in applications is a vector function which consists of a finite sum of periodic vector functions with periods which are incommensurable. There exists no common period; for instance the function

$$\sin t + \sin 2\pi t.$$

This function is not periodic but each of the two terms separately is periodic, periods $2\pi$ and 1.

We can formulate the following result.

**Theorem 11.2**

Consider the initial value problem

$$\dot{x} = \varepsilon f(t,x) + \varepsilon^2 g(t,x,\varepsilon)\,, \quad x(0) = x_0$$

with $x, x_0 \in D \subset \mathbf{R}^n$, $t \geq 0$. We assume that

a. the vector functions $f, g$ and $\partial f / dx$ are defined, continuous and bounded by a constant (independent of $\varepsilon$) in $[0, \infty) \times D$;

b. $g$ is Lipschitz-continuous in $x$ for $x \in D$;

c. $f(t, x) = \sum_{i=1}^{N} f_i(t, x)$ with $N$ fixed; $f_i(t, x)$ is $T_i$-periodic in $t$, $i = 1, \ldots, n$ with the $T_i$ constants independent of $\varepsilon$;

d. $y(t)$ is the solution of the initial value problem

$$\dot{y} = \varepsilon \sum_{i=1}^{N} \frac{1}{T_i} \int_0^{T_i} f_i(t, y) dt \ , \ y(0) = x_0$$

and $y(t)$ is contained in an interior subset of $D$. Then $x(t) - y(t) = O(\varepsilon)$ on the time-scale $1/\varepsilon$.

**Proof**
In the proof of theorem 11.1 we replace $f^0(y)$, $u(t, y)$ etc. by finite sums. □

Theorem 11.2 is a simple extension of theorem 11.1; the reason to formulate the theorem separately is that its conditions are often met in applications. In replacing the finite sum for $f(t, x)$ by a convergent series of periodic functions, we do encounter a problem of a different nature. One of the causes of this difference is that the expression $u(t, x)$, used in the near-identity transformation, need not be bounded anymore.
We now present a theorem for the case that the average of $f(t, x)$ exists in a more general sense.

**Theorem 11.3**
Consider the initial value problem

$$\dot{x} = \varepsilon f(t, x) + \varepsilon^2 g(t, x, \varepsilon) \ , \ x(0) = x_0$$

with $x, x_0 \in D \subset \mathbb{R}^n, t \geq 0$. We assume that

a. the vector functions $f, g$ and $\partial f / dx$ are defined, continuous and bounded by a constant (independent of $\varepsilon$) in $[0, \infty] \times D$;

b. $g$ is Lipschitz-continuous in $x$ for $x \in D$;

c. the average $f^0(x)$ of $f(t, x)$ exists where

$$f^0(x) = \lim_{T \to \infty} \frac{1}{T} \int_0^T f(t, x) dt;$$

d. $y(t)$ is the solution of the initial value problem

$$\dot{y} = \varepsilon f^\circ(y), \; y(0) = x_0.$$

Then $x(t) - y(t) = O(\delta(\varepsilon))$ on the time-scale $1/\varepsilon$ with

$$\delta(\varepsilon) = \sup_{x \in D} \sup_{0 \le t \le C} \varepsilon \| \int_0^t [f(s, x) - f^\circ(x)]ds \|.$$

**Proof**
See Sanders and Verhulst (1985), sections 3.3 and 3.4. $\qquad\qquad\qquad\square$

If we replace assumption a in theorem 11.3 by the requirement that $f$ is Lipschitz-continuous in $x$, we have the weaker estimate

$$x(t) - y(t) = O(\delta^{1/2}(\varepsilon)) \text{ on the time-scale } 1/\varepsilon.$$

We shall now discuss a simple example in which the theorem and the part played by the error $\delta(\varepsilon)$ is demonstrated.

**Example (increasing friction) 11.5**
Consider the oscillator with damping described by

$$(11.28) \qquad\qquad\qquad \ddot{x} + \varepsilon f(t)\dot{x} + x = 0.$$

We assume that for $t \ge 0$, the function $f(t)$ increases monotonically; $f(0) = 1$, $\lim_{t \to \infty} f(t) = 2$. It follows from theorem 6.3 that $(0,0)$ is asymptotically stable. As an illustration we consider two cases for $f(t)$:

$$f_1(t) = 2 - e^{-t}$$

and

$$f_2(t) = 2 - (1 + t)^{-1}.$$

In the first case the frictions increases more rapidly. With 11.14 we have in amplitude-phase variables

$$(11.29) \qquad\qquad \begin{aligned} \dot{r} &= -\varepsilon r \sin^2(t + \psi)f(t) \\ \dot{\psi} &= -\varepsilon \sin(t + \psi)\cos(t + \psi)f(t). \end{aligned}$$

We compute the averages of the righthandsides in the sense of theorem 11.3. It is easy to see that both in the case of the choice $f = f_1(t)$ and $f = f_2(t)$ we find the averaged equations

$$\dot{r}_a = -\varepsilon r_a , \; \dot{\psi}_a = 0.$$

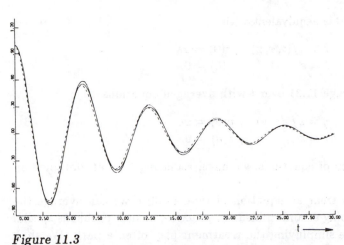

**Figure 11.3**
*Solutions of equation 11.28 with $f(t) = f_1(t)$ and $f(t) = f_2(t)$; the asymptotic approximation $x_a(t)$ has been indicated by $----$*

So in both cases we have the approximation

$$x_a(t) = r_0 e^{-\varepsilon t} \cos(t + \psi(0))$$

with $x(t) - x_a(t) = O(\delta(\varepsilon))$ on the time-scale $1/\varepsilon$.

This is a somewhat unexpected result as it means that right from the beginning the oscillator behaves approximately as if the friction has already reached its maximal value. We compute the error $\delta(\varepsilon)$ in both cases. By elementary estimates of the integrals we find:

$f(t) = 2 - e^{-t}$ yields $\delta(\varepsilon) = O(\varepsilon)$;

$f(t) = 2 - (1 + t)^{-1}$ yields $\delta(\varepsilon) = O(\varepsilon \mid ln\varepsilon \mid)$.

In figure 11.3 the solutions in the cases $f = f_1(t)$ and $f = f_2(t)$ have been obtained by numerical integration; they can be identified by realising that the solution corresponding with $f = f_1(t)$ experiences the strongest damping. In the case of $f = f_1(t)$, the approximation $x_a(t)$ is characterised by a better asymptotic error estimate than in the case $f = f_2(t)$. Also, it fits the numerical result slightly better.

## 11.5   Adiabatic invariants

In technical and physical applications mechanical systems arise which are modeled by differential equations with coefficients which are changing slowly with time. Simple examples are the pendulum with variable length (Einstein pendulum) and systems in which the mass is varying slowly, for instance a rocket using up fuel. Suppose that the equations of the problem can be formulated as

(11.30)                              $\dot{x} = \varepsilon f(t, \varepsilon t, x) \,, \; x(0) = x_0,$

with $x \in \mathbb{R}^n, t \geq 0$. In this case it is easy to obtain the Lagrange standard form and often we can apply one of the theorems 11.1-3. We introduce $\tau = \varepsilon t$ and the initial value problem 11.30 is aequivalent with

(11.31)
$$\dot{x} = \varepsilon f(t, \tau, x) \quad , x(0) = x_0$$
$$\dot{\tau} = \varepsilon \qquad\qquad , \tau(0) = 0.$$

Suppose that we can average 11.31 over $t$ with averaged equations

$$\dot{y} = \varepsilon f^\circ(\tau, y) \quad , \ y(0) = x_0$$
$$\dot{\tau} = \varepsilon \qquad\qquad , \ \tau(0) = 0.$$

If we can solve this system of equations, we have, replacing $\tau$ by $\varepsilon t$, obtained an approximation of $x(t)$.

We conclude that we can treat an equation of type 11.30 if we can average the equation over $t$ while keeping $\varepsilon t$ and $x$ fixed. In practice, in such problems special transformations which are simplifying the treatment play often a part. We shall demonstrate this in a simple case.

## Example 11.6
Consider a linear oscillator with slowly varying frequency.

(11.32)
$$\ddot{x} + \omega^2(\varepsilon t)x = 0.$$

Put $\dot{x} = \omega(\varepsilon t)y$ and transform $x, y \to r, \phi$ by

(11.33)
$$x = r\sin\phi, y = r\cos\phi.$$

Putting again $\tau = \varepsilon t$, equation 11.32 is aequivalent with the system

(11.34)
$$\dot{r} = -\varepsilon \frac{1}{\omega(\tau)} \frac{d\omega}{d\tau} r\cos^2\phi$$
$$\dot{\phi} = \omega(\tau) + \varepsilon \frac{1}{\omega(\tau)} \frac{d\omega}{d\tau} \sin\phi\cos\phi$$
$$\dot{\tau} = \varepsilon.$$

Now we add the assumption that for $t \geq 0$

$$0 < a < \omega(\tau) < b \ , \ |\frac{d\omega}{d\tau}| < c$$

with $a, b$ and $c$ constants, independent of $\varepsilon$. Note that the righthandside of system 11.34 is periodic in $\phi$ and as $\phi$ is a time-like variable (it is increasing monotonically with time) it makes sense to try averaging over $\phi$. The simplest way to do this is to replace the independent variable $t$ by $\phi$; a more general discussion will be given in the next section. System 11.34 becomes with $\phi$ as independent variable

(11.35)
$$\frac{dr}{d\phi} = -\varepsilon \frac{\frac{1}{\omega(\tau)}\frac{d\omega}{d\tau} r\cos^2\phi}{\omega(\tau) + \varepsilon\frac{1}{\omega(\tau)}\frac{d\omega}{d\tau}\sin\phi\cos\phi}$$
$$\frac{d\tau}{d\phi} = \varepsilon \frac{1}{\omega(\tau) + \varepsilon\frac{1}{\omega(\tau)}\frac{d\omega}{d\tau}\sin\phi\cos\phi}.$$

We can apply theorem 11.1 to this system. Averaging over $\phi$ produces

$$\frac{dr_a}{d\phi} = -\frac{\varepsilon}{2}\frac{r_a}{w^2(\tau_a)}\frac{dw}{d\tau_a}, \quad \frac{d\tau_a}{d\phi} = \frac{\varepsilon}{w(\tau_a)}.$$

From these equations we find

$$\frac{dr_a}{d\tau_a} = -\frac{r_a}{2w(\tau_a)}\frac{dw}{d\tau_a}$$

and after integration

$$w^{1/2}(\tau_a)r_a = C$$

with $C$ a constant determined by the initial values. In the original variables for equation 11.32 this means that

$$(11.36) \qquad w(\varepsilon t)x^2 + \frac{1}{w(\varepsilon t)}\dot{x}^2 = C^2 + O(\varepsilon) \text{ on the time-scale } 1/\varepsilon.$$

Here the time-like variable has been replaced again by $t$, the estimate still holds because of the assumptions for $w$.

In example 11.6 we have seen a very remarkable phenomenon. The oscillator corresponding with equation 11.32 has for $\varepsilon = 0$ the energy integral

$$E = \frac{1}{2}w^2(0)x^2 + \frac{1}{2}\dot{x}^2.$$

For $\varepsilon > 0$ the energy is not conserved but it follows from 11.36 that the quantity

$$\left(\frac{1}{2}w^2(\varepsilon t)x^2 + \frac{1}{2}\dot{x}^2\right)/w(\varepsilon t)$$

is conserved with accuracy $O(\varepsilon)$ on the time-scale $1/\varepsilon$. Such a quantity which has been conserved asymptotically while the coefficients are varying slowly with time, is called an *adiabatic invariant*.

## Definition
Consider the equation $\dot{x} = f(x, \varepsilon t)$ with $x \in \mathbb{R}^n$, $t \geq 0$. A function $I = I(x, \varepsilon t)$ is called an adiabatic invariant of the equation if

$$I(x, \varepsilon t) = I(x(0), 0) + o(1) \text{ on the time-scale } 1/\varepsilon.$$

The theory of adiabatic invariants has been developed in particular for Hamiltonian systems with coefficients which are slowly varying with time. In the last part of this section we shall sketch the ideas.
The starting point is a Hamilton function

$$H = H(p, q, \varepsilon t)$$

which induces the equations of motion

$$\dot{p} = -\frac{\partial H}{\partial q} \ , \ \dot{q} = \frac{\partial H}{\partial p} \ , \ p, q \in \mathbb{R}^n.$$

For $\tau = \varepsilon t$ fixed there exists a transformation $p, q \rightarrow I, \phi$ such that, in the new variables $I$, $\phi$, the system is again Hamiltonian with Hamilton function $h(I, \phi, \tau)$. If $n = 1$ and keeping $\tau$ fixed this transformation leads to simple equations :

$$\dot{I} = 0, \ \dot{\phi} = \omega(I, \tau).$$

$I$ and $\phi$ are called *action-angle coordinates*; in the original $p, q$-variables the system has the first integral $H$ ($\tau$ still fixed), in the new variables the action $I$ is a first integral.

## Remark

When introducing action-angle variables in the case $n > 1$, the equations of motion in general do not assume such a simple, integrable form. This is tied in with the generic non-integrability of Hamiltonian systems, see section 15.5.

The introduction of action-angle variables is usually carried out by employing a so-called *generating function* $S$ (see Arnold, 1976) which has the (rather artificial looking) properties

$$p = \frac{\partial S}{\partial q} \ , \phi = \frac{\partial S}{\partial I} \ ; \ S = S(I, q, \tau).$$

Replacing in the transformation $\tau$ by $\varepsilon t$ we have for $n = 1$ the system

$$\dot{I} = \varepsilon f(I, \phi, \varepsilon t); \qquad f \ = \ \partial^2 S / \partial \phi \partial \tau$$
$$\dot{\phi} = \omega(I, \varepsilon t) + \varepsilon g(I, \phi, \varepsilon(t)); \quad g \ = \ -\partial^2 S / \partial I \partial \tau.$$

The discussion of this system runs along the same lines as the discussion in example 11.6. We can use $\tau$ as a dependent variable and, if $\omega$ is bounded away from zero, we can average over $\phi$. If $\varepsilon = 0$, phase-space is two-dimensional and averaging over $\phi$ in the part of phase-space where the solutions are $2\pi$-periodic in $\phi$ produces

$$\int_0^{2\pi} f(I, \phi, \tau) d\phi = 0.$$

This is caused by the fact that $f$ is the derivative of a function which is periodic in $\phi$. So if $n = 1$, the action $I$ is an adiabatic invariant.

In the case $n > 1$, more than one degree of freedom, this procedure involves averaging over more angles; as is shown in the next sections, this is technically much more complicated and our knowledge of adiabatic invariance in such cases is still restriced. The statements in the literature concerning adiabatic invariants in systems with more than one degree of freedom have to be looked at with a more than usual critical eye.

## 11.6   Averaging over one angle, resonance manifolds

In a number of examples, for instance 11.6, we have seen that systems of equations arise which contain two types of dependent variables : $x \in \mathbb{R}^n$ and $\phi$ which represents an angular variable, i.e. $0 \leq \phi \leq 2\pi$ or $\phi \in S^1$ (the circle or 1-torus). This is quite natural in mechanical systems like nonlinear oscillators with or without coefficients which are slowly varying with time, gyroscopic systems and Hamiltonian systems. The equations are of the form

$$(11.37) \qquad \begin{aligned} \dot{x} &= \varepsilon X(\phi, x) + O(\varepsilon^2) \, , \, x \in D \subset \mathbb{R}^n \\ \dot{\phi} &= \Omega(x) + O(\varepsilon) \, , \phi \in S^1. \end{aligned}$$

with $X(\phi, x)$ periodic in $\phi$ and $t \geq 0$.
We have seen in example 11.6 that if $\Omega(x)$ is bounded away from zero, we can replace time $t$ by the time-like variable $\phi$ which yields

$$\frac{dx}{d\phi} = \varepsilon \frac{X(\phi, x)}{\Omega(x)} + O(\varepsilon^2)$$

after which we may apply theorem 11.1 while averaging over $\phi$.
We shall now present a theorem, the proof of which has the advantage that it can be generalized to the case of more than one angle.

**Theorem 11.4**
Consider system 11.37 with initial values $x(0) = x_0$, $\phi(0) = \phi_0$ and suppose that:

a. the righthand sides are $C^1$ in $D \times S^1$ ;

b. the solution of

$$\frac{dy}{dt} = \varepsilon X^0(y) \, , \, y(0) = x_0$$

with

$$X^0(y) = \int_{S^1} X(\phi, x) d\phi$$

is contained in an interior subset of $D$ in which $\Omega(x)$ is bounded away from zero by a constant independent of $\varepsilon$;

then $x(t) - y(t) = O(\varepsilon)$ on the time-scale $1/\varepsilon$.

**Proof**
We introduce the near-identity-transformation

$$x = z + \varepsilon u(\phi, z)$$

with

$$u(\phi, z) = \frac{1}{\Omega(z)} \int^{\phi} (X(\theta, z) - X^0(z)) d\theta.$$

System 11.37 becomes with this transformation

$$\dot{x} = \dot{z} + \varepsilon \frac{1}{\Omega(z)}(X(\phi, z) - X^0(z))\dot{\phi} + \varepsilon \frac{\partial u}{\partial z}\dot{z} = \varepsilon X(\phi, z + \varepsilon u(\phi, z)) + O(\varepsilon^2)$$

$$\dot{\phi} = \Omega(z + \varepsilon u(\phi, z)) + O(\varepsilon)$$

Expanding as in the proof of theorem 11.1 while using the smoothness of the vector functions we find

$$\begin{aligned} \dot{z} &= \varepsilon X^0(z) + O(\varepsilon^2) \\ \dot{\phi} &= \Omega(z) + O(\varepsilon). \end{aligned}$$

Note that $u(\phi, z)$ is uniformly bounded as long as assumption $b$ holds. On adding the initial value $z(0) = y(0)$, theorem 9.1 tells us that $y(t) - z(t) = O(\varepsilon)$ on the time-scale $1/\varepsilon$. Because of the near-identity-transformation we also have $x(t) - y(t) = O(\varepsilon)$ on the time-scale $1/\varepsilon$. □

## Remark 1

This first-order approximation of the solutions of system 11.37 doesnot produce an asymptotic approximation of $\phi(t)$ on the time-scale $1/\varepsilon$. To achieve this we have to compute a higher-order approximation, see Sanders and Verhulst (1985) chapter 5.

## Remark 2

Assumption $a$ can be relaxed somewhat; the reader can check this while referring to the proof of theorem 11.1.

What happens in regions where $\Omega(x) = 0$ or is near to zero? The set of points in $D$ where $\Omega(x) = 0$ will be called the *resonance manifold*. Let us start by examining an example.

## Example 11.7

Consider the system

$$\begin{aligned} \dot{x} &= \tfrac{1}{2}\varepsilon - \varepsilon \cos \phi &, x \in \mathbb{R} \\ \dot{\phi} &= x - 1 &, \phi \in S^1. \end{aligned}$$

In this problem $\Omega(x) = x - 1$ so the resonance manifold is given by $x = 1$. First we apply theorem 11.4 with the assumption that $x(0) = x_0$ is not near the resonance manifold. Averaging over $\phi$ produces

$$\dot{y} = \frac{1}{2}\varepsilon, \; y(0) = x_0$$

or $y(t) = x_0 + \frac{1}{2}\varepsilon t$; $x(t) - y(t) = O(\varepsilon)$ on the time-scale $1/\varepsilon$ as long as $y(t)$ doesnot enter a neighbourhood of the resonance manifold $(x = 1)$. Note that if $x_0 > 1$, the solution will stay away from the resonance manifold, if $x_0 < 1$, for instance $x_0 = \frac{1}{2}$, the solution will enter the neighbourhood of the resonance manifold and at this stage the approximation is no longer valid.

It is instructive to analyse what is going on near the resonance manifold. First we note that there are two critical points : $x = 1, \phi = \pi/3$ and $x = 1, \phi = 5\pi/3$; the first one is a saddle, the second one a centre point. Putting the initial values in one of these critical points we have clearly $x(t) - y(t) = \frac{1}{2}\varepsilon t$. Differentiating the equation for $\phi$ we derive easily

$$\ddot{\phi} + \varepsilon \cos \phi = \frac{1}{2}\varepsilon.$$

This equation has no attractors, so either a solution starts in or near the resonance manifold and remains there or it starts outside the resonance manifold and passes through this domain or it stays outside the resonance manifold altogether.

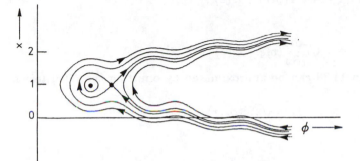

Figure 11.4. The $x, \phi$-phase-plane with the resonance manifold at $x = 1, \varepsilon = .1$. Solutions stay near the resonance manifold or they are passing through it

In the second case there is the interesting phenomenon that solutions which start initially close can be dispersed when passing through resonance. In figure 11.4 the $x, \phi$-phase-plane illustrates some of these phenomena. It is clear from the phase-plane that outside a neighbourhood of the resonance manifold the variable $\phi$ is time-like, while inside it is not.

To study the behaviour of the solutions in the vicinity of the resonance manifold we introduce local variables near the zeros of $\Omega(x)$. So consider again equation 11.37

$$\dot{x} = \varepsilon X(\phi, x) + O(\varepsilon^2) \quad , x \in D \subset \mathbb{R}^n$$
$$\dot{\phi} = \Omega(x) + O(\varepsilon) \quad\quad , \phi \in S^1$$

and suppose that $\Omega(r) = 0$ with $r \in \mathbb{R}^n$. We introduce

$$x = r + \delta(\varepsilon)\xi$$

with $\delta(\varepsilon) = o(1)$ as $\varepsilon \to 0$. Such a scaling is natural in boundary layer theory; for an extensive discussion of boundary layer variables see Eckhaus (1979). The order function $\delta(\varepsilon)$ will be determined by a certain balancing principle. We introduce the local variable $\xi$ in the equations and expand

$$
\begin{aligned}
\delta(\varepsilon)\dot{\xi} &= \varepsilon X(\phi, r + \delta(\varepsilon)\xi) + O(\varepsilon^2) = \varepsilon X(\phi, r) + O(\varepsilon\delta) + O(\varepsilon^2) \\
\dot{\phi} &= \Omega(r + \delta(\varepsilon)\xi) + O(\varepsilon) = \delta(\varepsilon)\tfrac{\partial\Omega}{\partial x}(r)\xi + O(\delta^2) + O(\varepsilon).
\end{aligned}
$$

The terms in the first and the second equations balance, i.e. they are of the same order if

$$
\delta(\varepsilon) = \sqrt{\varepsilon} \text{ and } \frac{\partial\Omega}{\partial x}(r) \neq 0.
$$

This determines the size of the boundary layer, i.e. the order of magnitude of the neighbourhood of the resonance manifold. The system becomes with this choice of $\delta(\varepsilon)$

$$
\dot{\xi} = \sqrt{\varepsilon}X(\phi, r) + O(\varepsilon) \;,\; \xi \in \mathbb{R}^n
$$

(11.38)

$$
\dot{\phi} = \sqrt{\varepsilon}\frac{\partial\Omega}{\partial x}(r)\xi + O(\varepsilon) \;,\; \phi \in S^1.
$$

The solutions of system 11.38 can be approximated by omitting the $O(\varepsilon)$ terms and solving

$$
\dot{\xi}_a = \sqrt{\varepsilon}X(\phi_a, r)
$$

(11.39)

$$
\dot{\phi}_a = \sqrt{\varepsilon}\frac{\partial\Omega}{\partial x}(r)\xi_a
$$

Theorem 9.1 tells us that, with appropriate initial values, the solutions of system 11.38 and system 11.39 are $\sqrt{\varepsilon}$-close on the time-scale $1/\sqrt{\varepsilon}$. System 11.39 is $(n+1)$-dimensional, the corresponding phase-flow is *volume-preserving* (lemma 2.4). This is surprising as the original system 11.37 may contain dissipation and forcing terms and has a quite general form. The phase-flow of system 11.39 can be characterised by a *two-dimensional* system; differentiate the second equation of 11.39 and we find after eliminating $\dot{\xi}_a$

(11.40)
$$
\ddot{\phi}_a - \varepsilon\frac{\partial\Omega}{\partial x}(r)X(\phi_a, r) = 0
$$

Equation 11.40 is a second-order equation; it is easy to write down a first integral (as for the mathematical pendulum) and critical points of the equation can only be saddles and centres.

This means of course that to describe correctly what is going on in the resonance manifold, we must compute a second-order approximation. In some problems we may find at second order that critical points which at first order are centre points, become at second order positive attractors. This opens the possibility that solutions of system 11.37 can be attracted into the boundary layer near the resonance manifold.

## 11.7  Averaging over more than one angle, an introduction

If more than one angle is present in the problem the set of resonance manifolds may be very much more complicated which causes the approximation theory to be correspondingly intricate. We discuss briefly an example with two angles.

**Example 11.8**
Consider the system

$$(11.41) \qquad \begin{aligned} \dot{x} &= \varepsilon X(\phi_1, \phi_2 . x) &&, x \in \mathbb{R} \\ \dot{\phi}_1 &= x &&, \phi_1 \in S^1 \\ \dot{\phi}_2 &= 1 &&, \phi_2 \in S^2. \end{aligned}$$

The initial values $x(0), \phi_1(0)$ and $\phi_2(0)$ are supposed to be given, $X$ is $2\pi$-periodic in $\phi_1$ and $\phi_2$. Averaging over the two angles means averaging over $S^1 \times S^1 = T^2$, the 2-torus. The theory of the preceding section suggests that we have to exclude regions where the righthand sides of the equations for the angles vanish, in this example the point $x = 0$ only, However, this is not sufficient. Consider the (complex) Fourier expansion of $X$

$$X(\phi_1, \phi_2, x) = \sum_{k,l=-\infty}^{\infty} c_{kl}(x) e^{i(k\phi_1 + l\phi_2)}$$

Averaging over the 2-torus produces

$$\frac{1}{(2\pi)^2} \int \int_{T^2} X(\phi_1, \phi_2, x) d\phi_1 d\phi_2 = c_{0\,0}(x)$$

provided that $k\phi_1 + l\phi_2 \neq 0$. This condition applies only if the coefficient $c_{kl}(x)$ is not identically zero. The quantity $k\phi_1 + l\phi_2$ can be zero or small for some interval of time in subsets of $\mathbb{R}$ where

$$k\dot{\phi}_1 + l\dot{\phi}_2 = 0$$

or, in the case of system 11.41,

$$kx + l1 = 0$$

Resonance manifolds in this example are given by the rationals $x = l/k$ where $k, l \in \mathbb{Z}$ are such that $c_{kl}(x)$ is not identically zero.
The result can be understood better on realising that one may replace $\phi_1$ and $\phi_2$ by two independent linear combination angles. If we take $\psi = k\phi_1 + l\phi_2$ as one of the combination angles, the equation for $\psi$ is varying slowly near the resonance manifold and we cannot average over $\psi$.

It is clear from example 11.8 that in the case of averaging over the $m$-torus the actual calculations and estimates will be much more difficult. We summarize some of the results which are known at present.

Consider the system with $m$ angles

(11.42)
$$\dot{x} = \varepsilon X(\phi, x) , x \in \mathbf{R}^n$$
$$\dot{\phi} = \Omega(x) \qquad , \phi \in T^m$$

where $\Omega(x) = (\Omega_1(x), \ldots, \Omega_m(x)), x = (x_1, \ldots, x_n), \phi = (\phi_1, \ldots, \phi_m)$. The resonance manifolds are given by

(11.43)
$$k_1\Omega_1(x) + \ldots + k_m\Omega_m(x) = 0, \quad x \in \mathbf{R}^n \text{ and}$$

$(k_1, \ldots, k_m) \in \mathbf{Z}^m$ such that the corresponding terms $c_{k_1 \ldots k_m}(x)$ in the Fourier expansion of $X(\phi, x)$ are not identically zero. In applications there is nearly always a finite number of such terms, or, if there is an infinite number, most coefficients are small and can be neglected.

Outside the set of resonance manifolds we average over the $m$ angles to obtain

$$\dot{y} = \varepsilon \int_{T^m} X(\phi, y)d\phi = \varepsilon c_{0 \ldots 0}(y)$$

where $c_{0 \ldots 0}$ indicates the Fouriercoefficient which remains. If $y(0) = x(0)$ we have again

$$x(t) - y(t) = O(\varepsilon) \text{ on the time-scale } 1/\varepsilon$$

as long as we do not enter the neighbourhood of a resonance manifold. The proof runs along the same lines as the proof of theorem 11.4.

One can analyse the flow near the resonance manifolds by introducing again local variables. Consider an isolated solution of equation 11.43 which corresponds with a $(n-1)$-dimensional manifold. To simplify the presentation, without reducing its generality, we assume that the coordinate system has been chosen such that the resonance manifold is given by

$$x_1 = 0.$$

The corresponding combination angle is

$$\psi = k_1\phi_1 + \ldots + k_m\phi_m.$$

We replace $\phi$ by $\psi$ and $(m-1)$ independent angles, say $\phi_2, \ldots, \phi_m$. The boundary layer near the resonance manifold can be described with the local variable $\xi$ where

$$x_1 = \delta(\varepsilon)\xi;$$

$\delta(\varepsilon) = o(1)$ and will be determined later. Putting $x = (x, \eta)$ with $\eta \in \mathbf{R}^{n-1}$, system 11.42 becomes in these new variables

$$\delta(\varepsilon)\dot{\xi} = \varepsilon X_1(\psi, \phi_2, \ldots, \phi_m, \delta(\varepsilon)\xi, \eta)$$
$$\dot{x}_i = \varepsilon X_i(\psi, \phi_2, \ldots, \phi_m, \delta(\varepsilon)\xi, \eta) , i = 2, \ldots, n$$
$$\dot{\psi} = \sum_{i=1}^{m} k_i\Omega_i(\delta(\varepsilon)\xi, \eta)$$
$$\dot{\phi}_i = \Omega_i(\delta(\varepsilon)\xi, \eta) , i = 2, \ldots, m.$$

Supposing that the vector functions are sufficiently smooth, we can expand to obtain

$$
\begin{aligned}
\delta(\varepsilon)\dot{\xi} &= \varepsilon X_1(\psi, \phi_2, \ldots, \phi_m, 0, \eta) + O(\varepsilon\delta(\varepsilon)) \\
\dot{x}_i &= \varepsilon X_i(\psi, \phi_2, \ldots, \phi_m, 0, \eta) + O(\varepsilon\delta(\varepsilon)), i = 2, \ldots, n \\
\dot{\psi} &= \delta(\varepsilon)\sum_{i=1}^{m} k_i \frac{\partial \Omega_i}{\partial x_1}(0, \eta)\xi + O(\delta^2(\varepsilon)) \\
\dot{\phi}_i &= \Omega_i(0, \eta) + O(\delta(\varepsilon)), i = 2, \ldots, m.
\end{aligned}
$$

As in the preceding section where we discussed the case of one angle, a balancing of terms arises if $\delta(\varepsilon) = \sqrt{\varepsilon}$. So in a $O(\sqrt{\varepsilon})$ neighbourhood of the resonance manifold the equations can be written as

(11.44)
$$
\begin{aligned}
\dot{\xi} &= \sqrt{\varepsilon}X_1(\psi, \phi_2, \ldots, \phi_m, 0, \eta) + O(\varepsilon) \\
\dot{\eta} &= O(\varepsilon) \\
\dot{\psi} &= \sqrt{\varepsilon}\sum_{i=1}^{m} k_i \frac{\partial \Omega_i}{\partial x_1}(0, \eta)\xi + O(\varepsilon) \\
\dot{\phi}_i &= \Omega_i(0, \eta) + O(\sqrt{\varepsilon}), i = 2, \ldots, m.
\end{aligned}
$$

This system is again in the form of system 11.42 but with small parameter $\sqrt{\varepsilon}$, $(n+1)$ slow variables $(\xi, \eta, \psi)$ and $(m-1)$ fast, time-like variables $\phi_2, \ldots, \phi_m$. Averaging over the $(m-1)$ fast variables is permitted and produces

(11.45)
$$
\begin{aligned}
\dot{\xi}_a &= \sqrt{\varepsilon}X_1^0(\psi_a, \eta_a) & , \xi_a(t_0) &= \xi(t_0) \\
\dot{\eta}_a &= 0 & , \eta_a(t_0) &= \eta(t_0) \\
\dot{\psi}_a &= \sqrt{\varepsilon}\sum_{i=1}^{m} k_i \frac{\partial \Omega_i}{\partial x_1}(0, \eta_a)\xi_a & , \psi_a(t_0) &= \psi(t_0).
\end{aligned}
$$

The solutions $\xi_a, \eta_a, \psi_a$ of system 11.45 with appropriate initial values approximate the solutions $\xi, \eta, \psi$ of system 11.44 with error $O(\sqrt{\varepsilon})$ on the time-scale $1/\sqrt{\varepsilon}$. This means that in this approximation the $(n-1)$ variables represented by $\eta$ are constant, the flow in the resonance manifold can be described by a system of dimension two. Differentiating the equation for $\psi_a$ we find

(11.46)
$$
\ddot{\psi}_a - \varepsilon\sum_{i=1}^{m} k_i \frac{\partial \Omega_i}{\partial x_1}(0, \eta_a)X_1^0(\psi_a, \eta_a) = 0.
$$

Although the system 11.42 , which is our starting-point, is $(n+m)$- dimensional, the equation describing the flow in an isolated resonance manifold is to order $\sqrt{\varepsilon}$ two-dimensional (as equation 11.40) and of pendulum-type. Non-degenerate critical points can only be saddles and centres. In most applications only one Fourier mode of the expansion for $X$ plays a part so that 11.46 very often becomes the mathematical pendulum equation with constant forcing

(11.47)
$$
\ddot{\psi}_a + \varepsilon\alpha(\eta_a)\cos\psi_a + \varepsilon\beta(\eta_a)\sin\psi_a = \varepsilon\gamma(\eta_a).
$$

It is clear that it is unavoidable in these problems to carry out a second- order calculation which involves the $(n+m)$ variables and which will generally change the centre points in the resonance manifold to attracting or repelling solutions. Such second-order calculations are carried out by Sanders and Verhulst (1985) and by van den Broek and Verhulst (1987). In this last reference one finds an application to a mechanical system consisting of a rotor-flywheel mounted on an elastic foundation. The reader may also consult a survey paper on systems with slowly varying coefficients, passage through a resonance manifold and related approximation techniques by Kevorkian (1987).

## 11.8  Periodic solutions

In section 10.4 we have seen that the Poincaré-Lindstedt method is not only a quantitative method but also leads, by the implicit function theorem, to the existence of periodic solutions. A similar result holds for the averaging method where again the implicit function theorem with an appropriate periodicity condition plays a part.
Consider again the equation

$$(11.48) \qquad \dot{x} = \varepsilon f(t,x) + \varepsilon^2 g(t,x,\varepsilon)$$

with $x \in D \subset \mathbf{R}^n, t \geq 0$. Moreover we assume that both $f(t,x)$ and $g(t,x,\varepsilon)$ are $T$-periodic in $t$. Separately we consider in $D$ the averaged equation

$$(11.49) \qquad \dot{y} = \varepsilon f^0(y).$$

Under certain conditions, equilibrium solutions of the averaged equation turn out to correspond with $T$-periodic solutions of equation 11.48.

**Theorem 11.5.**
Consider equation 11.48 and suppose that:

a.  the vector functions $f, g, \partial f/\partial x, \partial^2 f/\partial x^2$ and $\partial g/\partial x$ are defined, continous and bounded by a constant $M$ (independent of $\varepsilon$) in $[0,\infty) \times D, 0 \leq \varepsilon \leq \varepsilon_0$;

b.  $f$ and $g$ are $T$-periodic in $t$ ($T$ independent of $\varepsilon$);

If $p$ is a critical point of the averaged equation 11.49 whereas

$$(11.50) \qquad |\partial f^0(y)/\partial y|_{y=p} \neq 0$$

then there exists a $T$-periodic solution $\phi(t,\varepsilon)$ of equation 11.48 which is close to $p$ such that

$$\lim_{\varepsilon \to 0} \phi(t,\varepsilon) = p.$$

## Proof

First we shall impose the periodicity condition after which we can apply the implicit function theorem. We transform $x \to z$ with the near-identity relation 11.22

$$x(t) = z(t) + \varepsilon u(t, z(t)).$$

The equation for $z$ becomes

(11.51) $$\dot{z} = \varepsilon f^0(z) + \varepsilon^2 R(t, z, \varepsilon).$$

Because    of the choice of $u(t, z(t))$, a $T$-periodic solution $z(t)$ produces a $T$-periodic solution $x(t)$. For $R$ we have the expression

$$R(t, z, \varepsilon) = \frac{\partial f}{\partial z}(t, z)u(t, z) - \frac{\partial u}{\partial z}(t, z)f^0(z) + g(t, z, 0) + O(\varepsilon);$$

this expression is $T$-periodic in $t$ and continuously differentiable with respect to $z$. Equation 11.51 is aequivalent with the integral equation

$$z(t) = z(0) + \varepsilon \int_0^t f^0(z(s))ds + \varepsilon^2 \int_0^t R(s, z(s), \varepsilon)ds.$$

The solution $z(t)$ is $T$-periodic if $z(t + T) = z(t)$ for all $t \geq 0$ which leads to the equation

(11.52) $$h(z(0), \varepsilon) = \int_0^T f^0(z(s))ds + \varepsilon \int_0^T R(s, z(s), \varepsilon)ds = 0.$$

It is clear that $h(p, 0) = 0$. With $\varepsilon$ in a neighbourhood of $\varepsilon = 0$, equation 11.52 has a unique solution $z(0)$ because of the assumption on the Jacobi determinant 11.50. If $\varepsilon \to 0$, then $z(0) \to p$. $\quad\square$

If we have concluded with theorem 11.5 that a periodic solution of equation 11.48 exists in a neighbourhood of $x = p$, we can often establish its stability in a simple way:

## Theorem 11.6

Consider equation 11.48 and suppose that the conditions of theorem 11.5 have been satisfied. If the eigenvalues of the critical point $y = p$ of the averaged equation 11.49 all have negative real parts, the corresponding periodic solution $\phi(t, \varepsilon)$ of equation 11.48 is asymptotically stable for $\varepsilon$ sufficiently small. If one of the eigenvalues has positive real part, $\phi(t, \varepsilon)$ is unstable.

## Proof

We shall linearise equation 11.48 in a neighbourhood of the periodic solution $\phi(t, \varepsilon)$ after which we apply the theory of chapter 7. After translating $x = z + \phi(t, \varepsilon)$, expanding with respect to $z$, omitting the nonlinear terms and renaming the dependent variable again $x$, we find a linear equation with $T$-periodic coefficients:

(11.53) $$\dot{x} = \varepsilon A(t, \varepsilon)x$$

with $A(t, \varepsilon) = \frac{\partial}{\partial z}[f(t, x) + \varepsilon g(t, x, \varepsilon)]_{x=\phi(t,\varepsilon)}$.
We introduce the $T$-periodic matrix

$$B(t) = \frac{\partial f}{\partial x}(t, p).$$

From theorem 11.5 we have $\lim_{\varepsilon \to 0} A(t, \varepsilon) = B(t)$. We shall also use the matrices

$$B^0 = \frac{1}{T} \int_0^T B(t) dt$$

and

$$C(t) = \int_0^t [B(s) - B^0] ds.$$

Note that $B^0$ is the matrix of the linearised averaged equation. The matrix $C(t)$ is $T$-periodic and it has average zero. The near-identity transformation $x \to y$ with

$$y = (I - \varepsilon C(t))x$$

yields

(11.54)
$$
\begin{aligned}
\dot{y} &= -\varepsilon \dot{C}(t)x + (I - \varepsilon C(t))\dot{x} \\
&= -\varepsilon B(t)x + \varepsilon B^0 x + (I - \varepsilon C(t))\varepsilon A(t, \varepsilon)x \\
&= [\varepsilon B^0 + \varepsilon(A(t, \varepsilon) - B(t)) - \varepsilon^2 C(t) A(t, \varepsilon)](I - \varepsilon C(t))^{-1} y \\
&= \varepsilon B^0 y + \varepsilon(A(t, \varepsilon) - B(t))y + \varepsilon^2 R(t, \varepsilon)y.
\end{aligned}
$$

$R(t, \varepsilon)$ is $T$-periodic and bounded; we note that $(A(t, \varepsilon) - B(t)) \to 0$ as $\varepsilon \to 0$ and also that the characteristic exponents (section 6.3) of equation 11.54 depend continuously on the small parameter $\varepsilon$. It follows that, for $\varepsilon$ sufficiently small, the sign of the real parts of the characteristic exponents is equal to the sign of the real parts of the eigenvalues of the matrix $B^0$. The same conclusion holds, using the near-identity transformation, for the characteristic exponents of equation 11.53. We now apply theorem 7.2 to conclude stability in the case of negative real parts. If at least one real part is positive, Floquet transformation and application of theorem 7.3 leads to instability. □

**Example** (autonomous equations) **11.9**
For equations of type $\ddot{x} + x = \varepsilon f(x, \dot{x})$, and for autonomous equations in general, we shall always find one characteristic exponent with zero real part when linearising near a periodic solution; cf. sections 5.4 and 6.3. In the case of second-order

equations, it is then convenient to introduce polar coordinates and to consider the angle as a time-like variable. For instance in the case of the van der Pol- equation

$$\ddot{x} + x = \varepsilon(1 - x^2)\dot{x}$$

we have in phase-amplitude coordinates $r, \psi$ (example 11.3)

$$\dot{r} = \varepsilon \sin(t + \psi)(1 - r^2 \cos^2(t + \psi))r \sin(t + \psi)$$

$$\dot{\psi} = \varepsilon \cos(t + \psi)(1 - r^2 \cos^2(t + \psi)) \sin(t + \psi).$$

We put $t + \psi = \theta$ to obtain

$$\frac{dr}{d\theta} = \frac{\dot{r}}{1 + \dot{\psi}} = \varepsilon \sin \theta (1 - r^2 \cos^2 \theta) r \sin \theta + \varepsilon^2 \ldots$$

and after averaging over $\theta$

$$\frac{dr_a}{d\theta} = \frac{1}{2}\varepsilon r_a (1 - \frac{1}{4}r_a^2).$$

This equation has a critical point $r_a = 2$, the corresponding eigenvalue of $B^0$ in theorem 11.6 is $-1$. Application of theorem 11.5 yields the existence of a $2\pi$-periodic solution in $\theta$, according to theorem 11.6 this solution is asymptotically stable. The reader should check that averaging over $t$ of the equations for $r$ and $\psi$ doesnot lead to application of theorems 11.5-6. In using the Poincaré-Lindstedt method, where one also expands the period in the case of autonomous equations, this difficulty doesnot arise (see example 10.1).

**Example (forced Duffing equation) 11.10**
Consider again the Duffing equation with forcing and damping from section 10.3

$$\ddot{x} + \varepsilon\mu\dot{x} + x - \varepsilon x^3 = \varepsilon h \cos \omega t$$

with constants $\mu \geq 0, h > 0$; we put $\omega^{-2} = 1 - \varepsilon\beta$. We are looking for solutions which are periodic with period $2\pi/\omega$ so it makes sense to transform the variable $t$. Putting $\omega t = s$, the equations becomes (cf. 10.17)

$$(11.55) \qquad \frac{d^2x}{ds^2} + x = \varepsilon\beta x - \varepsilon(1 - \varepsilon\beta)^{\frac{1}{2}}\mu\frac{dx}{ds} + \varepsilon(1 - \varepsilon\beta)x^3 +$$

$$+\varepsilon(1 - \varepsilon\beta)h \cos s.$$

Introduction of amplitude-phase variables (section 11.2) produces

$$\begin{aligned}
\frac{dr}{ds} &= -\varepsilon \sin(s + \psi)[\beta r \cos(s + \psi) + \mu r \sin(s + \psi) + \\
&\quad + r^3 \cos^3(s + \psi) + h \cos s] + O(\varepsilon^2) \\
\frac{d\psi}{ds} &= -\varepsilon \cos(s + \psi)[\beta \cos(s + \psi) + \mu \sin(s + \psi) + \\
&\quad + r^2 \cos^3(s + \psi) + \frac{h}{r} \cos s] + O(\varepsilon^2).
\end{aligned}$$

The righthand sides are $2\pi$-periodic in $s$ and averaging yields (see appendix 3)

(11.56)
$$\frac{dr_a}{ds} = -\tfrac{1}{2}\varepsilon\mu r_a - \tfrac{1}{2}\varepsilon h \sin\psi_a$$
$$\frac{d\psi_a}{ds} = -\tfrac{1}{2}\varepsilon\beta - \tfrac{3}{8}\varepsilon r_a^2 - \tfrac{1}{2}\varepsilon\frac{h}{r_a}\cos\psi_a.$$

Critical points of the averaged equations 11.56 satisfy the transcendental equations

$$\mu r_a = -h\sin\psi_a \ , \quad \beta + \frac{3}{4}r_a^2 = -\frac{h}{r_a}\cos\psi_a.$$

These critical points correspond with periodic solutions of equation 11.55 if

$$\mu\sin\psi_a + \beta\cos\psi_a + \frac{9}{4}r_a^2\cos\psi_a \neq 0.$$

In accordance with the proof of theorem 11.5, this calculation produces the same equations as in the Poincaré-Lindstedt method when imposing the periodicity condition; cf. equation 10.20.

Finally we use theorem 11.6 to study the stability of the periodic solutions. Linearisation of system 11.56 in a critical point $P$ produces the matrix

$$\begin{pmatrix} -\tfrac{1}{2}\varepsilon\mu & -\tfrac{1}{2}h\varepsilon\cos\psi_a \\ -\tfrac{3}{4}\varepsilon r_a + \tfrac{1}{2}\tfrac{h}{r_a^2}\varepsilon\cos\psi_a & \tfrac{1}{2}\tfrac{h}{r_a}\varepsilon\sin\psi_a \end{pmatrix} =$$

$$\begin{pmatrix} -\tfrac{1}{2}\varepsilon\mu & \tfrac{1}{2}\varepsilon(\beta r_a + \tfrac{3}{4}r_a^3) \\ -\tfrac{9}{8}\varepsilon r_a - \tfrac{\varepsilon}{2r_a}\beta & -\tfrac{1}{2}\varepsilon\mu \end{pmatrix}_P$$

The eigenvalues are

$$\lambda_{1,2} = -\frac{1}{2}\varepsilon\mu \pm \frac{1}{2}\varepsilon[-(\beta + \frac{9}{4}r_a^2)(\beta + \frac{3}{4}r_a^2)]^{\frac{1}{2}}.$$

If for instance $\mu > 0$, $\beta \geq 0$ we have asymptotic stability.

## 11.9 Exercises

11-1. The equation
$$\ddot{x} + x + \varepsilon x^3 = 0$$
with $\varepsilon$ a small positive parameter, has one critical point; the other solutions are all periodic.

   a. Show this.

   b. Construct a first-order approximation of the solutions for general (not $\varepsilon$-dependent) initial values.

11-2. Supposing that the solutions of the equation $\ddot{x} + p(t)\dot{x} + q(t)x = 0$ are $T$-periodic, produce the Lagrange standard form for the equation

$$\ddot{x} + p(t)\dot{x} + q(t)x = \varepsilon f(x, \dot{x}).$$

11-3. Consider an oscillator with nonlinear friction described by the equation

$$\ddot{x} + x = \varepsilon f(\dot{x})$$

with $f(\dot{x})$ a function which can be expanded in a Taylor series in a sufficiently large neighbourhood of $\dot{x} = 0$.

a. Transform the equation to the Lagrange standard form.

b. Choose initial values and approximate the solutions by averaging.

c. The solutions of the unperturbed problem ($\varepsilon = 0$) are isochronous and they have a constant amplitude.
Is this also the case if $\varepsilon > 0$?

d. Discuss the behaviour of the solutions if $f(\dot{x})$ is even.

11-4. Averaging has been developed originally for celestial mechanics. The computations are usually cumbersome, we include here a relatively simple problem. A satelite moves in the atmosphere of a spherically symmetric, homogeneous planet. The forces are gravitation and the resisting force of the atmosphere; putting the origin of coordinates at the centre of the planet, the motion of the satellite is described by the equation

$$\frac{d^2\vec{r}}{dt^2} = -\frac{\vec{r}}{r^3} - \varepsilon\frac{d}{dt}\vec{r}$$

where we normalised the gravitational constant, $\vec{r} = (x, y, z)$, $r = (x^2 + y^2 + z^2)^{1/2}$. The resisting force (friction) has been taken linear.

a. Show that for given initial conditions, the motion takes place in planes through $(0, 0, 0)$. (note: we are discussing here motion in physical space, not phase-space).

b. From now on we take $z = 0$; because of $a$, this is no restriction of generality. Introduce polar coordinates $x = r\cos\theta, y = r\sin\theta$ to find the equations

$$\ddot{r} - r\dot{\theta}^2 = -\frac{1}{r^2} - \varepsilon\dot{r}$$
$$2\dot{r}\dot{\theta} + r\ddot{\theta} = -\varepsilon r\dot{\theta}$$

c. Integrate the second equation once to find

$$\ddot{r} = c^2 \frac{e^{-2\varepsilon t}}{r^3} - \frac{1}{r^2} - \varepsilon\dot{r}$$
$$r^2\dot{\theta} = ce^{-\varepsilon t}$$

The constant $c$ is called the initial angular momentum of the satellite.

d. A large part of work in celestial mechanics is devoted to putting equations in a tractable form. In this case we transform $\rho = 1/r$ and we use $\theta$ as a time-like variable. Show that we find the system

$$\frac{d^2\rho}{d\theta^2} + \rho = u$$
$$\frac{du}{d\theta} = 2\varepsilon\frac{u^{3/2}}{\rho^2}$$

and $u = \frac{1}{c^2}e^{2\varepsilon t}$.

e. Obtain the Lagrange standard form by transforming $(\rho, \frac{d\rho}{d\theta}) \rightarrow (a, b)$ with

$$\rho = u + a\cos\theta + b\sin\theta$$
$$\frac{d\rho}{d\theta} = -a\sin\theta + b\cos\theta.$$

f. Apply averaging and give the approximations for $\rho(\theta)$ and $r(t)$ with initially a circular orbit: at $t = 0$ $\theta(0) = 0$, $r(0) = c^2$, $\dot{r}(0) = 0$. Discuss the asymptotic validity.

11-5. The first-order equation $\dot{x} = \varepsilon(a + \sin t - x), a \in \mathbb{R}$ has a periodic solution; show this. Is the solution stable?

11-6. We are interested in the periodic solutions and their stability of the system $\dot{x} = \varepsilon f(t, x)$ with $x \in \mathbb{R}^3$ and

$$f(t, x) = \begin{matrix} x_1(2x_2\cos^2 t + 2x_3\sin^2 t - 2) \\ x_2(2x_3\cos^2 t + 2x_1\sin^2 t - 2) \\ x_3(2x_1\cos^2 t + 2x_2\sin^2 t - 2) \end{matrix}$$

Use averaging to find $2\pi$-periodic solutions.

11-7. The equation $\ddot{x} + x = \varepsilon(1 - ax^2 - b\dot{x}^2)\dot{x}$ with $a$ and $b$ positive constants has a periodic solution. Prove this and construct an asymptotic approximation of this solution.

11-8. In exercise 7.4 we studied a model for 2 competing species, in exercise 9.2 we have obtained a naïve perturbation expansion. We shall now discuss this problem by using nonlinear variation of constants and general averaging. The equations are

$$\dot{x} = x - x^2 - \varepsilon xy \quad , x \geq 0$$
$$\dot{y} = y - y^2 - \varepsilon xy \quad , y \geq 0$$

with positive initial values $x(0) = x_0, y(0) = y_0$.

a. Transform the system to the Lagrange standard form.

b. Note that the averaged system does not exist.
Introduce the time-like variable $\tau = e^t$ and perform averaging in $\tau$.

c. Discuss the asymptotic validity of the approximation. Is the approximation valid for all time?

11-9. As in exercise 11.4 we can study various perturbations of the gravitational two-body problem. Limiting the velocities to the velocity of light in the frame-work of relativity one can formulate equations of motion based on the equations of geodesics in Schwarzschild space-time. Using again $\rho = 1/r$ and the position angle $\theta$ one can derive the equation

$$\frac{d^2\rho}{d\theta^2} + \rho = \mu + \varepsilon\rho^2$$

with $\mu$ a constant; $\varepsilon = 0$ corresponds with the two-body problem with Newtonian gravitation.

a. Derive a Lagrange standard form.

b. Compute an approximation of the solutions by averaging.

11-10. Consider the system

$$\dot{x} = y + \varepsilon(x^2 \sin 2t - \sin 2t)$$
$$\dot{y} = -4x.$$

a. Compute a Lagrange standard form for this system and the corresponding averaged equations.

b. What is the evidence which one can derive from the averaged system on the existence and stability of periodic solutions?

11-11. We are studying an oscillator with linear damping and a variable frequency given by

$$\ddot{x} + \omega^2 x = -\tfrac{1}{2}\varepsilon\dot{x}$$
$$\dot{\omega} = \varepsilon(x^2 - \dot{x}^2), \omega(0) \geq \tfrac{1}{2}.$$

a. Put the system in the standard form 11.37 by the transformation $x, \dot{x} \rightarrow r, \phi$: $x = r\cos\phi, \dot{x} = -\omega r \sin\phi$.

b. Determine an $O(\varepsilon)$ approximtion of $r(t)$ and $\omega(t)$; suppose that $r(0)$ and $\omega(0)$ are given.

c. Show that we can apply theorems 11.5-6 to system 11.37 to obtain periodic solutions from the critical points of the averaged equations.

d. Determine the critical point(s) of the system obtained in b. Does a periodic solution in $\phi$ exist?

# 12  Relaxation oscillations

## 12.1  Introduction

Relaxation oscillations are periodic phenomena with very special features during a period. The characteristics can be illustrated by the following mechanical system.

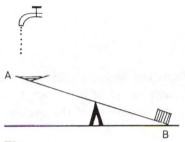

Figure 12.1

Consider a seesaw with at one side $(A)$ a container in which water can be held. If the container is empty, the other side $(B)$ of the seesaw touches the horizontal plane. From a tap water is dripping into the container and at a certain height of the waterlevel, point $B$ rises and point $A$ will touch the horizontal plane. At this moment the container empties itself, the seesaw returns quickly to its original position and the process starts again. On plotting the distance of the point $A$ to the horizontal plane, we obtain a graph as depicted in figure 12.2. This is a simple example of a relaxation oscillation.

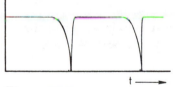

Figure 12.2

Such an oscillation is characterised by intervals of time in which very little happens, followed by short intervals of time in which notable changes take place. Apart from the field of classical mechanics, relaxation oscillations arise in many parts of physics, the engineering sciences and economy. For instance certain pulsating motions of stars have been associated with relaxation oscillations; this also applies to geophysical phenomena as the periodic bursts of steam in geysers and the sudden displacement of tectonic plates which are causing earthquakes. In mathematical biology one studies applications to periodic phenomena like the heartbeat, the respiratory movements of the lungs and other cyclic phenomena.

In this chapter we restrict ourselves to a first introduction to the mathematics of relaxation oscillations. One of the reasons for this restriction is that the quantitative treatment of this type of oscillations is technically complicated. A systematic exposition of methods and applications can be found in the monograph by Grasman (1987).

## 12.2 The van der Pol-equation

Again we consider the equation

(12.1)
$$\ddot{x} + x = \mu(1 - x^2)\dot{x}, \ \mu > 0.$$

In section 4.4 we have shown that this equation admits one periodic solution. We shall study now the behaviour of this solution for large values of $\mu : \mu \gg 1$. A numerical integration leads to an approximation of the behaviour of the periodic solution with time as given in figure 12.3;

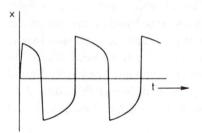

Figure 12.3

it is characterised by fast changes of the position $x$ near certain values of time $t$. We transform equation 12.1 in a way, first introduced by Liénard. We put

$$f(x) = -x + \frac{1}{3}x^3$$

and transform $x, \dot{x} \rightarrow x, y$ by

$$\mu y = \dot{x} + \mu f(x) \ , \ x = x.$$

The van der Pol-equation 12.1 becomes

(12.2)
$$\begin{aligned} \dot{x} &= \mu(y - f(x)) \\ \dot{y} &= -x/\mu \end{aligned}$$

The equation for the phase-flow in the $x, y$-plane becomes

(12.3)
$$(y - f(x))\frac{dy}{dx} = -\frac{x}{\mu^2}.$$

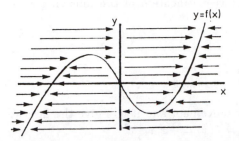

*Figure 12.4*

The closed curve, corresponding with the limit cycle in the phase-plane, can be described intuitively as follows. As $\mu \gg 1$, the righthand side of 12.3 is small. This suggests that the orbits can be described by the equation

$$(y - f(x))\frac{dy}{dx} = 0$$

which implies that either $y = f(x)$ or $y$ is constant. The correctness of this intuitive reasoning becomes clear on adding an analysis of the vector field of the flow induced by system 12.2. We find that outside a neighbourhood of the curve given by $y = f(x)$, the variable $x(t)$ changes quickly; the variable $y(t)$ is always changing slowly. This produces the flow field indicated in figure 12.4.

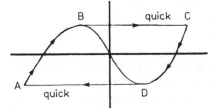

*Figure 12.5*

We can repeat now the application of the Poincaré-Bendixson theorem in section

4.4 for the case $\mu \gg 1$. We conclude that the limit cycle must be located in a neighbourhood of the curve sketched in figure 12.5.

A clear picture of the limit cycle has emerged in this way. To describe the quantitative behaviour of the periodic solution with time in terms of the parameter $\mu$, we need asymptotic methods from the theory of singular perturbations which are beyond the scope of this book; see Grasman (1987). If we use such a quantitative approximation method we also find a rigorous justification of the following estimate of the relaxation period $T$.

It follows from equation 12.2 that

$$
\begin{aligned}
T &= -\mu \int_{ABCDA} \frac{dy}{x} \\
&= -2\mu \int_{AB} \frac{dy}{x} - 2\mu \int_{BC} \frac{dy}{x}.
\end{aligned}
$$

The first integral corresponds with the slow movement and it will give the largest contribution to the period. We find with $y = f(x) = -x + \frac{1}{3}x^3$

$$
-2\mu \int_{AB} \frac{dy}{x} = -2\mu \int_{-2}^{-1} \frac{-1+x^2}{x} dy = (3 - 2\log 2)\mu.
$$

For the second integral we find with equation 12.3

$$
-2\mu \int_{BC} \frac{dy}{x} = \frac{2}{\mu} \int_{BC} \frac{x}{y - f(x)} dx
$$

which shows that this integral is smaller indeed in terms of $\mu$ apart from a contribution from the corner points $B$ and $C$. One can show that for the period $T$ of the periodic solution of the van der Pol-equation the following estimate holds

$$
T = (3 - 2\log 2)\mu + O(\mu^{-1/3}) \text{ as } \mu \to \infty.
$$

## 12.3 The Volterra-Lotka equations

We consider again the equations from the examples 2.7 and 2.15, a conservative system which models the development of two species, prey and predator:

$$
\begin{aligned}
\dot{x} &= ax - bxy \\
\dot{y} &= bxy - cy
\end{aligned}
$$
(12.4)

with $x, y \geq 0$ and positive parameters $a, b$ and $c$. The solutions for which $x(0), y(0) > 0$ are all periodic and the orbits are located around the critical point $(c/b, a/b)$ in the phase-plane. We assume now that the birth rate $a$ of a prey is much smaller than the death rate $c$ of the predator:

$$
\frac{a}{c} = \varepsilon.
$$

It is convenient to scale $p = \frac{d}{c}x, r = \frac{b}{a}y, \tau = at$. Replacing $x, y, t$ by $p, r, \tau$ and indicating the derivative with respect to $\tau$ by $'$, the Volterra-Lotka equations become

(12.5)
$$\begin{aligned} p' &= p(1-r) \\ \varepsilon r' &= r(-1+p). \end{aligned}$$

The critical point which is the centre of the closed orbits in the $p, r$- phase-plane is $(1, 1)$; the integral which we found in example 2.15 looks like

(12.6)
$$p - lnp + \varepsilon(r - lnr) = C.$$

As in the van der Pol-equation, the solution consists again of different parts. The periodic function $p(\tau)$ assumes extreme values if $r = 1$; these values can be obtained from the integral 12.6, they satisfy

$$p - lnp = C - \varepsilon.$$

The periodic function $r(\tau)$ on the other hand assumes extreme values if $p = 1$; from the integral 12.6 we find that they satisfy

$$\varepsilon(r - lnr) = C - 1.$$

The constant $C$ is determined by the initial values; $C - 1 > 0$. The extreme values of $r(\tau)$ are approximately of the order exp.$(-(C-1)/\varepsilon)$ and $(C-1)/\varepsilon$, so they are respectively exponentially small and of the order $1/\varepsilon$.

It follows from the second equation of system 12.5 that if $\varepsilon$ tends to zero, there are three possibilities: $r = 0$, $p = 1$ or $r(\tau)$ is varying quickly $(r' = O(1/\varepsilon))$. In the case that $r = 0$, the first equation produces

$$p(\tau) = p(0)e^{\tau}.$$

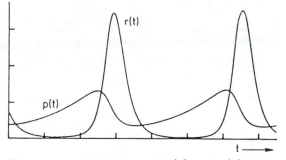

Figure 12.6. Behaviour of $p(\tau)$ and $r(\tau)$ for $\varepsilon = .1, C = .6+$ log 2

The number of preys increases exponentially. However, there is a maximal value of $p(\tau)$ where $r = 1$. So, if $p(t)$, which is increasing, enters a neighbourhood of this maximal value, $r(\tau)$ must increase quickly from a neighbourhood of 0 to 1. After $p(\tau)$ takes its maximal value, $r(\tau)$ increases from 1 to its maximal value and decreases after this (the reader should check this).

The interpretation is as follows. The number of preys is increasing when there are only a few predators; when the number of preys approaches its maximal value, suddenly the number of predators increases explosively at the cost of the preys. The number of preys decreases so that also the number of predators has to decrease. After this, the process can start all over again.

The equations of Volterra and Lotka are probably too simple to model population dynamics in real-life situations. On the other hand the equations represent a first, crude model for two species living together with a certain interaction.

One may compare the modelling situation with the part played by the harmonic oscillator in mechanics. Harmonic oscillators are nowhere to be found in nature but harmonic oscillation displays some basic phenomena which are helping us to understand more complicated real-life oscillations.

A classical illustration of two species polulation dynamics is derived from the trading figures of the Hudson Bay Company during the period 1845-1935.

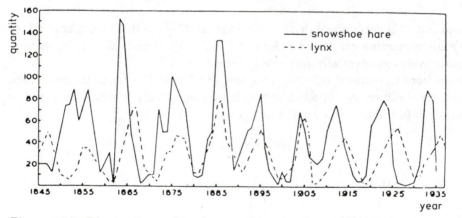

*Figure 12.7. Fluctuations of trade quantities (in thousands) of the Canadian lynx and the snowshoe hare*

# 13 Bifurcation theory

## 13.1 Introduction

In most examples of the preceding chapters, the equations which we have studied are containing parameters. For different values of these parameters, the behaviour of the solutions can be qualitatively very different. Consider for instance equation 7.12 in example 7.3 (population dynamics). When passing certain critical values of the parameters, a saddle changes into a stable node. The van der Pol-equation which we have used many times, for instance in example 5.1, illustrates another phenomenon. If the parameter $\mu$ in this equation equals zero, all solutions are periodic, the origin of the phase-plane is a centre point. If the parameter is positive with $0 < \mu < 1$, the origin is an unstable focus and there exists an asymptotically stable periodic solution, corresponding with a limit cycle around the origin. Another important illustration of the part played by parameters is the forced Duffing-equation in section 10.3 and example 11.8.

In this chapter we shall discuss changes of the nature of critical points and branching of solutions when a parameters passes a certain value; all this is called bifurcation theory. The foundations of the theory has been laid by Poincaré who studied branching of solutions in the three-body problem in celestial mechanics and bifurcation, i.e. splitting into two parts, of rotating fluid masses when the rotational velocity reached a certain value.

**Example 13.1**
Consider the equation

(13.1) $$\dot{x} = \mu x - x^2$$

The trivial solution $x = 0$ is an equilibrium solution of equation 13.2. Another equilibrium solution is $x = \mu$. these solutions coalesce if $\mu = 0$, at the value $\mu = 0$ both for positive and for negative values of $\mu$ a nontrivial solution branches off $x = 0$. The equation has a bifurcation at $\mu = 0$. the reader should sketch the solutions of the equation as a function of time for $\mu < 0, \mu = 0$ and $\mu > 0$. In passing the value $\mu = 0$ an exchange of stability of the equilibrium solutions $x = 0$ and $x = \mu$ takes place. This is illustrated in the so-called bifurcation diagram 13.1 which gives the equilibrium solutions as a function of the bifurcation parameter $\mu$.

In the example above we can predict the possibility of the existence of a branching or bifurcation point by the implicit function theorem. For the solutions $x$ of the

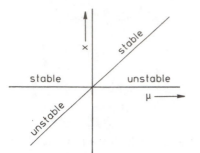

*Figure 13.1. Bifurcation diagram*

equation

$$F(\mu, x) = \mu x - x^2 = 0$$

exist and are unique if $\partial F/\partial x \neq 0$, i.e. $\mu - 2x \neq 0$. It is clear that the value $\mu = 0, x = 0$ where no uniqueness is guaranteed, is a good candidate for bifurcation.

Regarding critical points of differential equations we shall generally consider equations or a system of equations like

(13.2) $$F(\mu, x) = 0$$

with $\mu \in \mathbf{R}^m, x \in \mathbf{R}^n$. We are interested in the question whether a solution of equation 13.1 can bifurcate at certain values of the parameters $\mu = (\mu_1, \ldots, \mu_m)$. By translation we can assume without loss of generality that we are studying bifurcation of the trivial solution $x = 0$. So

$$F(\mu, 0) = 0.$$

**Definition**

Consider equation 13.1 with $F(\mu, 0) = 0$ (solution $x = 0$). The value of the parameter $\mu = \mu_c$ is called bifurcation value if there exists a nontrivial solution in each neighbourhood of $(\mu_c, o)$ in $\mathbf{R}^m \times \mathbf{R}^n$.

Another explicit calculation to illustrate this.

**Example 13.2**

Consider the equation

(13.3) $$\dot{x} = 1 - 2(1 + \mu)x + x^2.$$

The equation $1 - 2(1+\mu)x + x^2 = 0$ has unique solutions $x(\mu)$ if $-2(1+\mu) + 2x \neq 0$ or if $x \neq 1 + \mu$. Equation 13.3 has the equilibrium solutions $1 + \mu \pm (2\mu + \mu^2)^{\frac{1}{2}}$ for $\mu \leq -2$ and $\mu \geq 0$. Bifurcation can take place if $1 + \mu = 1 + \mu \pm (2\mu + \mu^2)^{1/2}$, i.e. for $\mu = 0$ and $\mu = -2$.

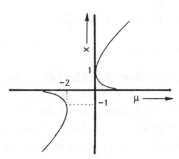

*Figure 13.2. Bifurcation diagram of example 13.2*

Note that both in example 13.1 and in example 13.2 we are considering a quadratic first order equation, in which the linear term depends linearly on $\mu$. The bifurcation diagram however, is in both cases very different. the similarity of equations 13.2 and 13.3 is very superficial which becomes clear on translating the equilibrium solutions of equation 13.3 to the origin (put $x = y + 1 + \mu + (2\mu + \mu^2)^{\frac{1}{2}}$ etc.).

Suppose now that we are interested in the bifurcations of the equilibrium solutions of the equation

$$(13.4) \qquad \dot{x} = A(\mu)x + f(\mu, x)$$

with $\mu \in \mathbf{R}^m, x \in \mathbf{R}^n, \partial f / \partial x \to 0$ as $\|x\| \to 0$. With regards to applications it would be useful if we would have at our disposal a classification of possible bifurcation diagrams. There are many results if the dimensions are low, $m = 1, n = 1$ and $n = 2$. There are some results for $n \geq 3$, but there are many new types of bifurcations at higher dimension and we are far removed from having a survey or a complete classification.

In a systematic study of bifurcation phenomena it is useful to transform the differential equations to a standard form. This process we call *normalisation*. One of the normalisation techniques has been discussed in section 11.3 where we introduced near-identity averaging transformation. We shall consider again near-identity transformations.

## 13.2  Normalisation

In this section we consider equations of the form

$$(13.5) \qquad \dot{x} = Ax + f(x)$$

with $A$ a constant $n \times n$-matrix; $f(x)$ can be expanded in homogeneous vector polynomials which start with degree 2. On writing this expansion as $f(x) = f_2(x) +$

$f_3(x) + \ldots$ the vector polynomial $f_m(x), m \geq 2$, contains terms of the form

$$x_1^{m_1} x_2^{m_2} \ldots x_n^{m_n} , \ m_1 + m_2 + \ldots + m_n = m.$$

If $\lambda$ is a constant we have $f_m(\lambda x) = \lambda^m f_m(x)$. If we are interested in the behaviour of the solutions in a neighbourhood of the critical point $x = 0$, it is useful to introduce near-identity transformations which simplify the vector function $f(x)$. Even better: we would like to find smooth transformations which turn equation 13.5 into a linear equation. Linearisation by transformation is in most important cases not possible, simplification of the equation is nearly always a possibility. As an introduction we consider the simple case of a one- dimensional equation.

### Example 13.3

Consider the equation

(13.6) $$\dot{x} = \lambda x + a_2 x^2 + a_3 x^3 + \ldots$$

with $\lambda \neq 0, x \in \mathbf{R}$. We introduce the near-identity transformation in the form of a series

(13.7) $$x = y + \alpha_2 y^2 + \alpha_3 y^3 + \ldots$$

where we will try to determine the coefficients $\alpha_2, \alpha_3, \ldots$ such that the equation for $y$ is linear. If we are successful in this, the transformation 13.7 represents a formal expansion with respect to $y$. Maybe it is convergent for $y = 0$ only, we have to check this. Differentiating 13.7 and substituting into equation 13.6 produces

$$\dot{y}(1 + 2\alpha_2 y + 3\alpha_3 y^2 + \ldots) = \lambda y + (\lambda \alpha_2 + a_2)y^2 + (\lambda \alpha_3 + 2a_2\alpha_2 + a_3)y^3 + \ldots$$

Dividing by the coefficient of $\dot{y}$ we have

$$\dot{y} = \lambda y + (a_2 - \lambda \alpha_2)y^2 + (a_3 + 2\lambda \alpha_2^2 - 2\lambda \alpha_3)y^3 + \ldots$$

Requiring the coefficients of $y^2$ and $y^3$ to vanish we find

$$\alpha_2 = \frac{a_2}{\lambda} , \ \alpha_3 = \frac{a_2^2}{\lambda^2} + \frac{a_3}{2\lambda}.$$

By this choice of $\alpha_2$ and $\alpha_3$, equation 13.6 is normalised to degree three.

If the dimension of the equation is higher than one, the theory becomes more complicated. Equation 13.5 in the form

$$\dot{x} = Ax + f_2(x) + f_3(x) + \ldots$$

will be transformed by (cf. 13.7)

(13.8) $$x = y + h(y)$$

with $h(y)$ consisting of a, probably infinite, sum of homogeneous vector polynomials $h_m(y)$ with $m \geq 2$. So we can write transformation 13.8 as

$$x = y + h_2(y) + h_3(y) + \ldots$$

We would like to determine $h(y)$ such that $\dot{y} = Ay$. Substitution of 13.8 into 13.5
produces

$$\begin{aligned}\dot{x} = \dot{y} + \tfrac{\partial h}{\partial y}\dot{y} &= (I + \tfrac{\partial h}{\partial y})\dot{y} \\ &= A(y + h(y)) + f(y + h(y)).\end{aligned}$$

Inversion of $I + \partial h/\partial y$ in a neighbourhood of $y = 0$ yields

(13.9) $$\dot{y} = (I + \frac{\partial h}{\partial y})^{-1}[Ay + Ah(y) + f(y + h(y))].$$

We start with removing the quadratic terms in 13.9. Expansion of $h$ and $f$ yields
the equation

$$\frac{\partial h_2}{\partial y} Ay - Ah_2 = f_2(y).$$

This is the so-called *homology* equation for $h_2$. It is easy to see that, on requiring
that the terms of degree $m$ vanish, we have the homology equation

(13.10) $$\frac{\partial h_m}{\partial y} Ay - Ah_m = g_m(y) \ , \ m \geq 2.$$

For $m > 2$, the righthand sides $g_m(y)$ can be expressed in terms of the solutions (if
they exist) of the homology equation to degree $m - 1$. In considering the solvability
of the homology equation 13.10, we observe that the lefthand side is linear in $h_m$.
The linear mapping $L_A$, sometimes called ad $L$,

$$L_A(h) = \frac{\partial h}{\partial y} Ay - Ah(y)$$

carries homogeneous vector polynomials over in vector polynomials of the same
degree. If the set of eigenvalues of $L_A$ doesnot contain zero, $L_A$ is invertible
and equation 13.10 can be solved. For simplicity's sake we assume now that all
eigenvalues $\lambda_1, \ldots, \lambda_n$ of the matrix $A$ are different and that $A$ is in diagonal form.
Written out in conponents $h_m = (h_{m1}, \ldots, h_{mn})$ we find for 13.10

(13.11) $$\sum_{j=1}^{n} \frac{\partial h_{mi}}{\partial y_j} \lambda_j y_j - \lambda_i h_{mi}(y) = g_{mi}(y), i = 1, \ldots, n, \ m \geq 2.$$

The terms in $h_{mi}$ are all of the form

$$a y_1^{m_1} y_2^{m_2} \ldots y_n^{m_n} = a y^{(m)}$$

where $y^{(m)}$ is a shorthand notation and $m = m_1 + m_2 + \ldots + m_n$, $a$ is a constant. The eigenvectors of $A$ are $e_i$, the eigenvectors of $L_A$ are $y^{(m)} e_i$ with eigenvalues

$$(13.12) \qquad \sum_{j=1}^{n} m_j \lambda_j - \lambda_i \ , \ i = 1, \ldots, n.$$

If an eigenvalue of $L_A$ is zero, we call this *resonance*. If there is no resonance, equations 13.11 can be solved and the nonlinear terms in equation 13.9 can be removed.

**Definition**
The eigenvalues $\lambda_1, \ldots, \lambda_n$ of the matrix $A$ are *resonant* if for $i \in \{1, 2, \ldots, n\}$ one has

$$\lambda_i = \sum_{j=1}^{n} m_j \lambda_j$$

with $m_j \in \{0\} \cup \mathbb{N}$ and $m = m_1 + m_2 + \ldots + m_n \geq 2$.

If the eigenvalues $\lambda_1, \ldots, \lambda_n$ of the matrix $A$ are non-resonant, equation 13.5 can be put into linear form by transformation 13.8. Here, 13.8 is determined in the form of a series, the convergence of which one still has to study. This is why transformation 13.8 is called formal.
We summarize the results.

**Theorem 13.1 (Poincaré)**
If the eigenvalues of the matrix $A$ are non-resonant, equation 13.5 $\dot{x} = Ax + f_2(x) + \ldots$ can be transformed into the linear equation $\dot{y} = Ay$ by the formal transformation 13.8 $x = y + h_2(y) + \ldots$

Guckenheimer and Holmes (1983), section 3.3, give an explicit calculation for the important case $n = 2$, $\lambda_1 = i$, $\lambda_2 = -i$ which covers the cases of the perturbed harmonic oscillator like the van der Pol-equation. The eigenvalues of $L_A$ are with 13.12

$$m_1 i - m_2 i \mp i.$$

For $m_1 + m_2 = 2$, there is no resonace, so all quadratic terms can be removed. Now we can try to remove the cubic terms. Calculating the eigenvalues 13.11 of $L_A$ in the case $m = 3$ we find resonance. Some cubic terms can be removed, some will remain; see also example 13.5.

*In practice there is nearly always resonance or the eigenvalues are approximately resonant.*

Suppose we find for example no resonance for $m = 2, \ldots, r - 1$; for $m = r$ we find zero eigenvalues 13.12 of $L_A$. All terms of degree $2, 3, \ldots$ until degree $r - 1$ can be removed from equation 13.5 by transformation. For $m = r$, we still solve the homology equations 13.11 in the cases where the eigenvalues of $L_A$ are nonvanishing. The resonant terms of degree $r$ remain. These resonant terms donot influence the form of the homology equations, so we can continue the transformations for $m > r$. Finally we are left with a nonlinear equation for $y$ with resonant terms only. This process of removing non-resonant terms will be called normalisation. The names of Poincaré and Dulac are often mentioned in this context.

## Example 13.4
Consider the system

$$\dot{x}_1 = 2x_1 + a_1 x_1^2 + a_2 x_1 x_2 + a_3 x_2^2 + \ldots$$
$$\dot{x}_2 = x_2 + b_1 x_1^2 + b_2 x_1 x_2 + b_3 x_2^2 + \ldots$$

where the dots represent polynomials of degree three and higher. The eigenvalues of the linear part are $\lambda_1 = 2, \lambda_2 = 1$ and we consider the possibility of resonance by satisfying the relations

$$2m_1 + m_2 = 2 \text{ or } 2m_1 + m_2 = 1$$

with $m_1 + m_2 \geq 2$. It is a simple puzzle to find that the only possibility is $m_1 = 0, m_2 = 2$. This means that the system can put in the following normal form:

$$\dot{y}_1 = 2y_1 + cy_2^2$$
$$\dot{y}_2 = y_2.$$

This is a considerable simplification.

## Example 13.5
Consider the system describing perturbed harmonic oscillation

$$\dot{x}_1 = x_2 + \ldots$$
$$\dot{x}_2 = -x_1 + \ldots$$

where the dots represent polynomials of degree two and higher. The eigenvalues of the linear part are $\lambda_1 = i$ and $\lambda_2 = -i$. We have resonance if

$$m_1 i - m_2 i = \pm i \quad \text{or}$$

$$m_1 - m_2 = \pm 1 \text{ for } m_1 + m_2 \geq 2.$$

This means that all polynomials of even degree can be removed by transformation. An infinite number of terms of odd degree will remain in the normal form of the equations as for instance the resonant terms $y_1^2 y_2, y_1 y_2^2, y_1^3 y_2^2$ etc.

**Remarks**

1. If the matrix $A$ has multiple eigenvalues, the treatment given here carries through as the reader may check. But multiple eigenvalues do produce more resonances.

2. If one of the eigenvalues is zero, we have resonance. If for instance $\lambda_1 = 0$, we can take $m_1 = m, m_2 = \ldots = m_n = 0$ to satisfy the resonance relation. Later in this chapter we shall see that eigenvalues zero play an important part in bifurcation theory.

In example 13.4 it is possible to remove all nonlinear terms with the exception of a finite number of resonant terms. In example 13.5 the eigenvalues of the linear part are such that after normalisation an infinite number of resonant terms are present. The occurrence of a finite or infinite number of resonant terms can easily be predicted. Consider again the matrix $A$ in equation 13.5 $\dot{x} = Ax + f(x)$ which we are intending to normalise near the critical point $x = 0$. The $n$ eigenvalues $\lambda_1, \ldots, \lambda_n$ of the matrix $A$ are located in the complex plane $\mathbb{C}$.

**Theorem 13.2**

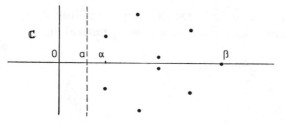

*Figure 13.3. The $n$ eigenvalues of matrix $A$ separated from zero by a straight line*

Consider equation 13.5 $\dot{x} = Ax + f(x)$. If all the eigenvalues of $A$ are lying either to the right or to the left of the imaginary axis in $\mathbb{C}$, the equation can be reduced to a polynomial normal form by a formal transformation of variables (13.8).

The assumption on the location of the eigenvalues can be put in a more geometric way by requiring that zero is not contained in the convex hull of the collection of eigenvalues. In the proof the number of eigenvalues being finite is essential.

**Proof**

If the eigenvalues are non-resonant, theorem 13.1 applies and theorem 13.2 follows immediately. Suppose now that there are resonances; a line parallel to the imaginary axis, intersecting the real axis at $a$, separates the collection of $n$ eigenvalues from zero. Take $a > 0$ (the proof with $a < 0$ runs in an analogous way) and put

$$\inf_{i=1...n} Re\lambda_i = \alpha > a, \quad \sup_{i=1...n} Re\lambda_i = \beta.$$

We choose $m = \sum_{j=1}^{n} m_j$ such that $m\alpha > \beta$; then the resonance relation $\lambda_i = \sum_{j=1}^{n} m_j\lambda_j$ cannot be satisfied for this value of $m$.

The implication is that for $m$ sufficiently large no resonant terms are encountered any more: the resulting normal form of the equation is polynomial. □

In this section we have introduced important formal transformations without bothering about the qualitative aspects of the calculations and results. In section 3.3 on the other hand, we have formulated some theorems on the relation between linear and nonlinear equations in a neighbourhood of a critical point with real parts of the eigenvalues non-vanishing. In section 3.3 we also remarked that the phase-flows near a critical point of the linear and the nonlinear equation in these cases are homeomorphic and in general not diffeomorphic. If the righthand side of the nonlinear equation is analytic, there is generally not an analytic transformation connecting linear and nonlinear equation. This last point is illustrated by example 13.4. Here the critical point $(0,0)$ is, after linearisation, a negatively attracting node. There exists generally not even a formal series transforming the nonlinear equations into the linear one.

The reader should consult the book by Arnold (1983) for more qualitative and quantitative results and references.

## 13.3 Averaging and normalisation

On introducing the averaging method in section 11.3, we remarked that the averaging transformation is an example of normalisation. We consider this remark again for averaging in the periodic case.

Consider again equation 13.5 in the form

$$\dot{x} = Ax + f_2(x) + f_3(x) + \ldots$$

with $f_m(x)$ homogeneous vector polynomials of degree $m$. The matrix $A$ is supposed to be in diagonal form and contains only purely imaginary eigenvalues $\lambda_j = w_j i, j = 1,\ldots,n$; the eigenvalues may be multiple. Expressing that we are studying a neighbourhood of $x = 0$ we scale $x = \varepsilon\bar{x}$ with $\varepsilon$ a small parameter. Equation 13.5 becomes

$$\varepsilon\dot{\bar{x}} = \varepsilon A\bar{x} + \varepsilon^2 f_2(\bar{x}) + \varepsilon^3 f_3(\bar{x}) + \ldots$$

Dividing by $\varepsilon$ and omitting the bars produces

(13.13) $$\dot{x} = Ax + \varepsilon f_2(x) + \varepsilon^2 \ldots$$

We use variation of parameters

$$x = e^{At}z$$

or, written out in components

$$x_1 = e^{w_1 it}z_1, \ldots, x_n = e^{w_n it}z_n.$$

The equation for $z$ is

(13.14) $$\dot{z} = \varepsilon e^{-At}f_2(e^{At}z) + \varepsilon^2 \ldots$$

For each component, the first (quadratic) term in the righthand side of the equation consists of a sum of expressions like

$$\varepsilon c e^{-w_j it}e^{m_1 w_1 it}e^{m_2 w_2 it}\ldots e^{m_n w_n it}$$

with $m_1 + m_2 + \ldots + m_n = 2$; so $m_k = 0, 1$ or $2$ for $k = 1, 2, \ldots, n$. The average of the righthand side over $t$ is zero unless we have for certain values of the parameters

$$\sum_{k=1}^{n} m_k w_k - w_j = 0.$$

This is precisely the case of resonance, defined in the preceding section. The resonant terms which remain after averaging are the same as the ones which remain in applying the method of Poincaré and Dulac. In the notation of chapter 11 we have the following near-identity transformation. Consider equation 13.14 in the form

(13.15) $$\dot{x} = \varepsilon f(t, x) + \varepsilon^2 g(t, x) + \varepsilon^3 \ldots$$

The righthand side of 13.15 is $T$-periodic, the average of $f(t, x)$ over $t$ is $f^0(x)$. We transform

(13.16) $$x(t) = z(t) + \varepsilon u^1(t, z(t))$$

with $u^1(t, z) = \int_0^t [f(s, z) - f^0(z)]ds$. Substituting into equation 13.15 we find for $z$ the equation

(13.17) $$\dot{z} = \varepsilon f^0(z) + \varepsilon^2 \left(\frac{\partial f(t, z)}{\partial z}u^1(t, z) - \frac{\partial u^1(t, z)}{\partial z}f^0(z) + \right.$$

$$\left. + g(t, z)\right) + \varepsilon^3 \ldots$$

The equation has been normalised to order $\varepsilon$; this means that the explicit time-dependence of equation 13.15 has been removed to $O(\varepsilon)$ terms. We can repeat the procedure to remove the time-dependence of the terms of order $\varepsilon^2$.

**Remark**
Normalisation á la Poincaré as in section 13.2, averaging and some other computational procedures can be put in a more general framework of normalisation; the

reader may consult Sanders and Verhulst (1985), chapter 6. A brief account runs as follows.

Consider a space of sufficiently smooth vector functions $V$. We introduce the commutator [,] called Poisson bracket, which is an operator, mapping $V$ into itself. Suppose that $S$ and $P$ are operators, mapping $V$ into itself, then we have

$$[S, P] = SP - PS.$$

In this frame-work it is also convenient to introduce the operator ad $S : P \to [S, P]$. Now we will say that the operator $S + P$ is in normal form if $P \in \ker(\text{ad } S)$ or somewhat more general: if $P \in \ker(\text{ad}^k S)$ for some $k \in \mathbb{N}$

For instance in normalising 13.5 $\dot{x} = Ax + f(x)$ we have $S = A, P = \frac{\partial}{\partial x}$; we identify

$$L_A(h) = \text{ad } S(h).$$

To normalise in the sense of Poincaré and Dulac, we have to split $f(x)$ in a resonant and a non-resonant part.

In normalising equation 13.15 by averaging, the appropriate identification is $S = \frac{\partial}{\partial t}, P = \frac{\partial}{\partial x}$. We have

$$[S, P](g) = \frac{\partial}{\partial t}\frac{\partial}{\partial x}g - f\frac{\partial}{\partial x}\frac{\partial}{\partial t}g = \frac{\partial f}{\partial t}\frac{\partial g}{\partial x}.$$

$[S, P](g) = 0$ if $f$ doesnot depend explicitly on $t$. This is achieved by the averaging transformation which turns equation 13.15 into its normal form equation 13.17 to order $\varepsilon$.

## 13.4   Centre manifolds

In section 3.3 we studied the equation

$$\dot{x} = Ax + g(x)$$

in the case that the constant $n \times n$-matrix $A$ has eigenvalues with non-vanishing real part. The point $x = 0$ is a critical point which has stable and unstable manifolds which are tangent in $x = 0$ to the corresponding manifolds of the equation $\dot{y} = Ay$. The case of eigenvalues of the matrix $A$ which are purely imaginary or even zero shows behaviour which can be rather different from the full nonlinear equation as compared with the linearised one; see the examples 3.2 and 3.3, also again the van der Pol-equation in 2.9.

If, on linearisation, we find eigenvalues zero or purely imaginary, bifurcations may arise; we have seen some simple examples in section 13.1. To study these phenomena we shall extend first theorem 3.3.

**Theorem 13.3**

Consider the equation

(13.18) $$\dot{x} = Ax + f(x)$$

with $x \in \mathbf{R}^n$ and $A$ a constant $n \times n$-matrix; $x = 0$ is isolated critical point. The vector function $f(x)$ is $C^k$, $k \geq 2$, in a neighbourhood of $x = 0$ and

$$\lim_{\|x\|\to 0} \frac{\|f(x)\|}{\|x\|} = 0.$$

The stable and unstable manifolds of equation $\dot{y} = Ay$ are $E_s$ and $E_u$, the space of eigenvectors corresponding with eigenvalues with real part zero is $E_c$. There exist $C^k$ stable and unstable invariant manifolds $W_s$ and $W_u$, which are tangent to $E_s$ and $E_u$ in $x = 0$. There exists a $C^{k-1}$ invariant manifold $W_c$, the centre manifold, which is tangent to $E_c$ in $x = 0$; if $k = \infty$, then $W_c$ is in general $C^m$ with $m \leq \infty$.

**Proof**

See Marsden and McCracken (1976) or Carr (1981). □

So a centre manifold is characterised by the fact that $W_c$ is an invariant set which contains $x = 0$ and which is tangent to $E_c$ in $x = 0$.

**Example 13.6**

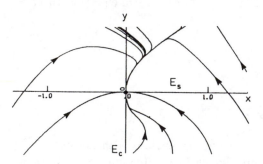

*Figure 13.4. $(0,0)$ is degenerate critical point*

Consider the system from example 3.3

$$\begin{aligned} \dot{x} &= -x + y^2 \\ \dot{y} &= -y^3 + x^2. \end{aligned}$$

The critical point $(0,0)$ has, after linearisation, eigenvalues $-1$ and $0$ so $(0,0)$ is degenerate. $E_s$ is the $x$-axis, $E_c$ the $y$-axis. According to theorem 13.3, $W_s$ and $W_c$ exist which are tangent to the $x$-axis and the $y$-axis at $(0,0)$. $W_s$ is $C^\infty$ and $W_c$ is $C^m$ with $m \leq \infty$.

Theorem 13.3 is very important in the study of systems with eigenvalues of which the real part is zero. If for instance there is no eigenvalue with positive real part, the phase-flow will show stable behaviour in a neighbourhood of $x = 0$, except perhaps in the centre manifold. In such cases we shall restrict ourselves to a study of the flow in this lower dimensional invariant manifold.

In the example above we have seen that the location of $W_c$ still has to be established more precisely. In example 13.6 $E_c$ is the $x$-axis and we cannot simply project the flow on $E_c$. Moreover we have to realise in approximation centre manifolds, that $W_s$ and $W_u$ are unique but $W_c$ is generally not unique. This is illustrated as follows.

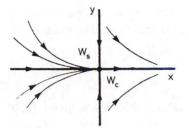

Figure 13.5. Saddle-node

**Example (A. Kelley) 13.7**

Consider the system

$$\dot{x} = x^2$$

$$\dot{y} = -y$$

In this case $E_s$ and $W_s$ coincide (positive and negative $y$- axis). $E_c$ coincides with the $x$-axis, in this example the $x$-axis is a part of $W_c$. The orbits in the phase-plane can be found by integration:

$$y(x) = Ce^{1/x}$$

where $C$ is determined by the initial values. To the left of the $y$-axis, the centre manifold consists of a infinite number of submanifolds which are all tangent to $E_c$. In this example $W_c$ is not analytic but $C^\infty$.

We shall now take a closer look at the use of centre manifolds when studying stability. Suppose that the matrix $A$ in equation 13.18 has eigenvalues with real part negative and zero. Assume for simplicity that the matrix is in diagonal form, so instead of 13.18 we may write

(13.19)
$$\begin{aligned}
\dot{x} &= Ax + f(x,y) \\
\dot{y} &= By + g(x,y)
\end{aligned}$$

with $x \in \mathbb{R}^n, y \in \mathbb{R}^m$; $A$ and $B$ are constant diagonal matrices, where $A$ has only eigenvalues with real part zero, $B$ has only eigenvalues with negative real part; $(x, y) = (0, 0)$ is isolated critical point. The functions $f$ and $g$ are $C^k, k \geq 2$, and have a Taylor expansion in a neighbourhood of $(0, 0)$ in $\mathbb{R}^{n+m}$ which contains no constant and no linear terms.

According to theorem 13.2 there exists $h : \mathbb{R}^n \to \mathbb{R}^m$, where $y(x) = h(x)$ represents the centre manifold of 13.19 in $(0, 0)$. The flow in the centre manifold determines the stability of the zero solution.

## Theorem 13.4

Consider equation 13.19 in which $A$ has only eigenvalues with real part zero, $B$ has only eigenvalues with negative real part. The flow in the centre manifold is determined by the $n$-dimensional equation

(13.20) $$\dot{u} = Au + f(u, h(u)).$$

If the solution $u = 0$ of equation 13.20 is stable (unstable), then the solution $(0, 0)$ of equation 13.19 is stable (unstable).

## Proof
See Carr (1981). $\qquad\qquad\square$

It is important to approximate $h(x)$ with regards to applications. Substituting $y = h(x)$ in the second equation of 13.19 produces

$$\frac{\partial h}{\partial x} \dot{x} = Bh(x) + g(x, h(x)).$$

Moreover we have in the centre manifold

$$\dot{x} = Ax + f(x, h(x)).$$

So we obtain a first order partial differential equation for $h$

(13.21) $$\frac{\partial h}{\partial x}(Ax + f(x, h)) - Bh - g(x, h) = 0$$

with $h(0) = 0$ and tangency condition $\frac{\partial h}{\partial x}(0) = 0$. We approximate $h(x)$ by substitution of a Taylor expansion into equation 13.21.

## Example (J. Carr) 13.8
Consider the system

(13.22) $$\begin{aligned} \dot{x} &= xy + ax^3 + bxy^2 \\ \dot{y} &= -y + cx^2 + dx^2y. \end{aligned}$$

According to theorem 13.3 the system has a stable manifold $W_s$ which is tangent to the $y$-axis at $(0,0)$. There exists a centre manifold $W_c$, given by $y = h(x)$, which is tangent to the $x$-axis. Equation 13.21 for $h$ takes the form

$$\frac{dh}{dx}(xh + ax^3 + bxh^2) = -h + cx^2 + dx^2h.$$

We substitute $h(x) = \alpha x^2 + \beta x^3 + \dots$ and equate coefficients of equal powers of $x$ on the left and the righthand side; we find $\alpha = c, \beta = 0$ so that $h(x) = cx^2 + O(x^4)$. Now we apply theorem 13.4.

The flow in the centre manifold is determined by equation 13.20:

$$\dot{u} = uh(u) + au^3 + buh^2(u)$$

and in a neighbourhood of $u = 0$

$$\dot{u} = (a + c)u^3 + O(u^5).$$

It follows from theorem 13.4 that the solution $(0,0)$ of system 13.22 is stable if $a + c < 0$, unstable if $a + c > 0$. If we would have $a + c = 0$, we have to compute higher order terms.

## 13.5 Bifurcation of equilibrium solutions and Hopf bifurcation

In the introduction 13.1 we saw a number of examples of bifurcations which arise when parameters are passing a certain value. Consider again equation 13.4

$$\dot{x} = A(\mu)x + f(\mu, x)$$

with $x \in \mathbb{R}^n, \mu$ a parameter in $\mathbb{R}$. We suspend this $n$-dimensional system in a $(n + 1)$-dimensional system by adding $\mu$ as new variable:

(13.23)
$$\begin{aligned} \dot{x} &= A(\mu)x + f(\mu, x) \\ \dot{\mu} &= 0 \end{aligned}$$

Suppose now that $\partial f/\partial x \to 0$ as $\|x\| \to 0$ and consider the possibility of bifurcation of the solution $x = 0$. If the matrix $A$ has $p$ eigenvalues with real part zero for a certain value of $\mu$, equation 13.23 has a $(p + 1)$-dimensional centre manifold $W_c$. To study $W_c$ we can formulate equation 13.20 for this case after which we can simplify the equation by normalisation according to section 13.2 or 13.3.

**Example 13.9**
Consider the system

(13.24)
$$\begin{aligned} \dot{x} &= \mu x - x^3 + xy \\ \dot{y} &= -y + y^2 - x^2. \end{aligned}$$

We are interested in bifurcations in a neighbourhood of $(0,0)$ for small values of $|\mu|$. We suspend equations 13.24 in the system

(13.25)
$$\begin{aligned} \dot{x} &= \mu x - x^3 + xy \\ \dot{y} &= -y + y^2 - x^2 \\ \dot{\mu} &= 0 \end{aligned}$$

The system, linearised in a neighbourhood of $(0,0,0)$, has the eigenvalues $(0,-1,0)$ so according to theorem 13.3, there exists a 2- dimensional centre manifold $y = h(x,\mu)$. Differentiating and by using system 13.25 we find

$$\begin{aligned} \tfrac{\partial h}{\partial x}(\mu x - x^3 + xh) &= -h + h^2 - x^2 \\ \dot{\mu} &= 0. \end{aligned}$$

Substituting for $h$ a Taylor expansion with respect to $x$ and $\mu$ we find

$$h(x,\mu) = -x^2 + \ldots$$

where the dots represent cubic and higher order terms in $x$ and $\mu$ (the reader should check this). In the centre manifold, the flow is determined by (cf. 13.20)

(13.26)
$$\begin{aligned} \dot{u} &= \mu u - 2u^3 + \ldots \\ \dot{\mu} &= 0 \end{aligned}$$

if $\mu \leq 0$, the solution $u = 0$ is asymptotically stable and it follows from theorem 13.4 that the same holds for the corresponding solution of system 13.25 and so for 13.24. If $\mu > 0$, the solution $u = 0$ is unstable. Moreover, for small values of $\mu > 0$ there exist two stable equilibrium solutions with order of magnitude $\pm(\mu/2)^{1/2}$. Below we shall indicate this bifurcation as "supercritical pitch-fork".

We shall apply these techniques systematically to equation 13.4 or system 13.23, where there is one eigenvalue of $A(\mu)$ which has real part zero. In this (simplest) case, the corresponding centre manifold of system 13.23 is 2-dimensional as in example 13.9. The flow in the centre manifold is described by an equation like 13.26, which we can simplify by normalisation. It turns out that the following cases arise frequently.

*The saddle-node bifurcation*
In the centre manifold the flow is described by

(13.27)
$$\begin{aligned} \dot{u} &= \mu - u^2 \\ \dot{\mu} &= 0 \end{aligned}$$

If $\mu < 0$, there exists no equilibrium solution. At $\mu = 0$ two equilibrium solutions branch off, of which one is stable and one unstable.

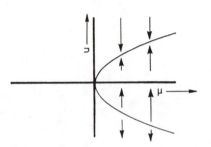

*Figure 13.6. Saddle-node bifurcation*

*The trans-critical bifurcation*
The flow is described by the equations

(13.28)
$$\begin{aligned} \dot{u} &= \mu u - u^2 \\ \dot{\mu} &= 0 \end{aligned}$$

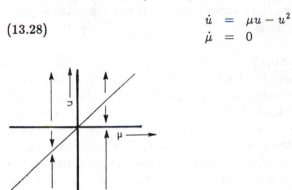

*Figure 13.7. Trans-critical bifurcation*

We have seen this case already in example 13.1. Apart from $(0,0)$ there are always two equilibrium solutions with an exchange of stability when passing $\mu = 0$.

*The pitch-fork bifurcation*
The flow is described by the equations

(13.29)
$$\begin{aligned} \dot{u} &= \mu u - u^3 \\ \dot{\mu} &= 0. \end{aligned}$$

If $\mu \leq 0$, there is one equilibrium solution, $u = 0$, which is stable. If $\mu > 0$, there are three equilibrium solutions, of which $u = 0$ is unstable, the two solutions which have branched off at $\mu = 0$ are stable. This is called the pitch-fork bifurcation and it is supercritical.
On replacing $-u^3$ by $+u^3$, the figure is reflected with respect to the $u$-axis; in this

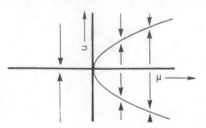

*Figure 13.8. Pitch-fork bifurcation, supercritical*

case the bifurcation is called subcritical.

**Remark**
In using these bifurcation diagrams one should keep in mind that they present a picture, based on local expansion. At some distance of the bifurcation value qualitatively new phenomena may arise.

*Bifurcation of periodic solutions (Hopf)*
This bifurcation was studied already by Poincaré, who, in his work on the gravitational three-body problem, obtained certain periodic solutions "of second type" (solutions of first type are obtaind by the continuation method of chapter 10). Later Hopf gave a more explicit discussion. One considers the case in which the matrix $A(\mu)$ in equation 13.4 or system 13.23 has two purely imaginary eigenvalues for a certain value of $\mu$, whereas the other eigenvalues all have a non-vanishing real part. We know this situation already from our study of the van der Pol-equation with $\mu$ near the value $\mu = 0$

$$\ddot{x} + x = \mu(1 - x^2)\dot{x}.$$

It is possible to put the parameter in the linear part only by the rescaling $u = \sqrt{\mu}x$. We find for $u$ the equation

$$\ddot{u} + u = (\mu - u^2)\dot{u}.$$

As we know, the equation for $x$ has for $\mu > 0$ one periodic solution which for small values of $\mu$ has the amplitude $2 + O(\mu)$. This means that for $u$ the periodic solution branches off with amplitude $2\sqrt{\mu} + O(\mu^{3/2})$. This behaviour of solutions of the van der Pol-equation is typical for Hopf-bifurcation.
More general consider

(13.30)
$$\begin{aligned} \dot{x} &= \mu x - \omega y + \dots \\ \dot{y} &= \omega x + \mu y + \dots \end{aligned}$$

where the dots represent quadratic and higher order terms, $\omega$ is fixed, $\omega \neq 0$. If $\mu = 0$ the eigenvalues of the linear part are purely imaginary.

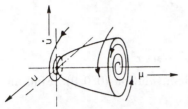

Figure 13.9. *Hopf-bifurcation of periodic solution*

Normalisation removes all quadratic terms (cf. example 13.5) and a number of cubic terms. To degree three, the normal form of equations 13.30 is

(13.31)
$$\dot{u} = d\mu u - (\omega + c\mu)v + a(u^2 + v^2)u - b(u^2 + v^2)v + \ldots$$
$$\dot{v} = (\omega + c\mu)u + d\mu v + b(u^2 + v^2)u + a(u^2 + v^2)v + \ldots$$

In polar coordinates the system looks better :

(13.32)
$$\dot{r} = (d\mu + ar^2)r + \ldots$$
$$\dot{\theta} = \omega + c\mu + br^2 + \ldots$$

At $\mu = 0$ we have a pitch-fork bifurcation for the amplitude ($r$) equation which corresponds with a Hopf-bifurcation for the full system. A periodic solution of equation 13.32 exists if $d \neq 0$ and $a \neq 0$ with amplitude $r = (-d\mu/a)^{1/2}$.
The proof of existence of this periodic solution of system 13.30 can be carried out as follows: first a reduction of system 13.23 to a system for the three-dimensional centre manifold (theorem 13.3 and 13.4), secondly normalisation and finally a proof in the spirit of theorem 11.4

## 13.6 Exercises

13-1. Consider the stability of $(0,0)$ in system

$$\dot{x} = xy$$
$$\dot{y} = -y + 3x^2.$$

Sketch the phase-flow near $(0,0)$.

13-2. Consider again the system in example 13.7

$$\dot{x} = x^2$$
$$\dot{y} = -y.$$

We have a non-unique $C^\infty$ centre manifold. Try to approximate the centre manifold(s) by a Taylor expansion.

13-3. The system

$$
\begin{aligned}
\dot{x} &= -y + xz - x^4 \\
\dot{y} &= x + yz + xyz \\
\dot{z} &= -z - (x^2 + y^2) + z^2 + \sin x^3
\end{aligned}
$$

is studied in a neighbourhood of the equilibrium solution $(0,0,0)$. Find an approximation for the centre manifold $W_c$.
Is $(0,0,0)$ stable?

13-4. Equation 13.3

$$\dot{x} = 1 - 2(1 + \mu)x + x^2$$

has been studied in example 13.2. Determine the stability of the equilibrium solutions.

13-5. Example 13.3 is rather trivial. Assuming that the righthand side of equation 13.6 is a convergent power series with respect to $x$ in a neighbourhood of $x = 0$, it is easy to see that 13.7 represents a linearising transformation of equation 13.6 which is a convergent series in the same neighbourhood. How?

13-6. Consider equation 13.5 in the case $n = 2$ near the critical point $x = 0$. The cases with infinitely many resonant terms have codimension 1 in a parameter space generated by the coefficients of the matrix $A$. For which non-degenerate critical points do they occur?

13-7. The Belgian scientists Prigogine and Lefever formulated a model for certain chemical reactions; the model is called the Brusselator. If there is no spatial diffusion the equations are

$$
\begin{aligned}
\dot{x} &= a - (b + 1)x + x^2 y \\
\dot{y} &= bx - x^2 y
\end{aligned}
$$

in which $x$ and $y$ are concentrations $(x, y \geq 0)$, $a$ and $b$ are positive parameters. Does the Brusselator admit the possibility of Hopf bifurcation?

13-8. In section 13.3 we have shown that averaging and normalisation involve the same resonance conditions. In this exercise we show that the approximation theory based on both techniques leads to asymptotically aequivalent results. Consider the equation

$$\ddot{x} + x = \varepsilon x^2$$

In example 11.1, omitting the friction term, we found $x(t) = r(0) \cos(t + \psi(0)) + O(\varepsilon)$ on the time-scale $1/\varepsilon$.

a. Write the equation as $\dot{x} = y, \dot{y} = -x + \varepsilon x^2$ and introduce the near-identity transformation 13.8 by

$$
\begin{aligned}
x &= u + \varepsilon(a_1 u^2 + a_2 uv + a_3 v^2) + \varepsilon^2 \dots \\
y &= v + \varepsilon(b_1 u^2 + b_2 uv + b_3 v^2) + \varepsilon^2 \dots
\end{aligned}
$$

Compute the normal form of the equations to order 2, i.e. to $O(\varepsilon)$.

b. Find an approximation of the solution with prescribed initial values and compare this with the result by averaging.

13-9. We are looking for bifurcation phenomena in the system

$$
\begin{aligned}
\dot{x} &= (1 + a^2)x + (2 - 6a)y + f(x,y) \\
\dot{y} &= -x \qquad\qquad -2y + g(x,y)
\end{aligned}
$$

with parameter $a$; $f(x,y)$ and $g(x,y)$ can be developed in a Taylor series near $(0,0)$ starting with quadratic terms.

a. For which values of $a$ does the Poincaré-Lyapunov theorem (7.1) quarantee asymptotic stability of $(0,0)$?

b. For which values of $a$ is it possible to have a bifurcation of $(0,0)$? Construct an example where this actually happens.

c. For which values of $a$ does a centre manifold exist?

d. For which values of $a$ is it possible to have Hopf bifurcation.

13-10. Consider the system

$$
\begin{aligned}
\dot{x} &= (x^2 + y^2 - 2)x \\
\dot{y} &= -y^3 + x
\end{aligned}
$$

Determine the stability of the stationary solution $(0,0)$.

# 14  Chaos

In this chapter we shall sketch a number of complicated phenomena which are tied in with the concept of "chaos" and "strange attraction". There are scientists who see also a relation with the phenomenon of turbulence in continuum mechanics but this interesting idea involves still many unsolved problems. We shall restrict ourselves to a discussion of two examples from the various domains where these phenomena have been found: autonomous differential equations with dimension $n \geq 3$, second-order forced differential equations like the forced van der Pol- or the forced Duffing equation and mappings of R into R, $R^2$ into $R^2$ etc.

In a subsequent chapter we shall meet the phenomenon of chaos again in the dynamics of Hamiltonian systems.

Good general introductions have been given by Schuster (1984), Bergé, Pomeau and Vidal (1984), Guckenheimer and Holmes (1983) and for mappings by Devaney (1986).

## 14.1  The Lorenz-equations

In 1963 there appeared a remarkable paper by the meteorologist E.N. Lorenz. In this paper the starting-point is the phenomenon of convection in the atmosphere of the Earth by heating from below and cooling from above. The model for this problem leads to the Boussinesq equations, which are nonlinear partial differential equations.

By considering three important modes in the system, mathematically speaking by projecting the infinite-dimensional space of solutions on a three- dimensional subspace, Lorenz derived the following system

$$(14.1) \qquad \begin{aligned} \dot{x} &= \sigma(y - x) \\ \dot{y} &= rx - y - xz \\ \dot{z} &= xy - bz \end{aligned}$$

in which $\sigma, r$ and $b$ are positive parameters. As $x, y$ and $z$ represent dominant modes of the convective flow, on physical grounds this projection seems to be a reasonable first approximation. We note however that on adding modes, i.e. by projection on a higher-dimensional subspace, we obtain very different results. However, even as the use of system 14.1 as a model for convection in the atmosphere is very much in dispute, the equations are mathematically very interesting. In the sequel we shall often take $\sigma = 10, b = 8/3, r$ can take various positive values. First we choose $r = 28$ and we construct a numerical approximation of a solution which starts in a neighbourhood of the unstable equilibrium solution $(0,0,0)$; it starts in $E_u$ so that the orbit follows the unstable manifold $W_u$ as well as possible.

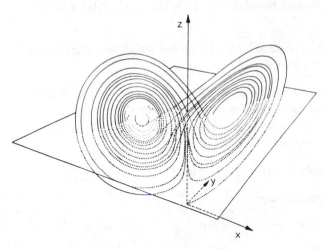

*Figure 14.1. Numerical approximation of one solution of the Lorenz- equations*

In figure 14.1 this orbit has been depicted; it turns out to have the following features:

a. The orbit is not closed.

b. The orbit doesnot represent a transition stage to well-known regular behaviour; the orbit continues to describe loops on the left and on the right without apparent regularity in the number of loops.

c. The subsequent number of loops on the left and on the right depend in a very sensitive way on the initial values. A small perturbation of the initial values produces another alternating series of loops.

d. Other initial values, also very different ones, produce roughly the same picture.

The numerical experiments suggest the existence of an attracting set with a dimension a little bigger than two, which, in the perspective of point $c$, has a complicated topological structure. For these new phenomena, Ruelle and Takens invented the name "strange attraction". Actually, there exist definitions of strange attraction, but until now it has been very difficult to apply these definitions in explicit examples.

A number of characteristics of the Lorenz-equations are easy to derive. We discuss them briefly, the reader should verify the details.

1. Equations 14.1 have the following reflection symmetry: replace $x, y, z$ by $-x, -y, z$ and the equations have the same form. It follows that each solution $(x(t), y(t), z(t))$ has a symmetric counter-part $(-x(t), -y(t), z(t))$ which is also a solution.

2. The $z$-axis, $x = y = 0$, is an invariant set. Solutions which are starting on the $z$-axis tend to $(0,0,0)$ for $t \to \infty$.

3. Each volume-element, mapped by the phase-flow into phase-space, shrinks. This can be seen by calculating the divergence of the vector function on the righthand side $\nabla.(\sigma y - \sigma x, rx - y - xz, xy - bz) = -(\sigma + 1 + b)$. With lemma 2.4 we have for the volume $v(t)$ of the element

$$\dot{v}(0) = -(\sigma + 1 + b)v(0).$$

As $t = 0$ is arbitrary, we clearly have

$$\dot{v} = -(\sigma + 1 + b)v$$

so that

$$v(t) = v(0) \exp. (-(\sigma + 1 + b)t).$$

A phase-flow with negative divergence is called *dissipative* (some authors are calling a system dissipative if the divergence doesnot vanish; this is confusing as example 2.15 of the Volterra-Lotka equations has shown). In a dissipative system there cannot exist critical points or periodic solutions which are a negative attractor. For, if this would be the case, in a neighbourhood of these solutions a volume-element would expand.

4. Equilibrium solutions.
For $0 < r < 1$ there is one equilibrium solution, corresponding with the critical point $(0,0,0)$. This solution is asymptotically stable. When $r$ passes the value 1, one eigenvalue becomes zero and we have a bifurcation. For $r > 1$ three critical points exist: $(0,0,0), (\pm\sqrt{b(r-1)}, \pm\sqrt{b(r-1)}, r-1)$. The point $(0,0,0)$ has three real eigenvalues, two are negative, one is positive, the last one corresponding with a one-dimensional unstable manifold. It follows that for $r > 1$, the solution $(0,0,0)$ is unstable.
The eigenvalues of the other two critical points which exist for $r > 1$ are satisfying the equation

$$\lambda^3 + \lambda^2(\sigma + b + 1) + \lambda b(\sigma + r) + 2\sigma b(r - 1) = 0.$$

It can be shown that if $1 < r < r_H$, the three roots of this cubic equation have all negative real parts; we have

$$r_H = \frac{\sigma(\sigma + b + 3)}{\sigma - b - 1}$$

If $r = r_H$ two of the eigenvalues are purely imaginary and we have Hopf bifurcation. This bifurcation turns out to be subcritical, i.e. for $r < r_H$ there exist two unstable periodic solutions corresponding with the two critical points; on passing the value $r = r_H$ these periodic solutions vanish.

If $r > r_H$, each of the two critical points has one negative (real) eigenvalue and two eigenvalues with real part positive so they correspond with unstable solutions. This is the case of figure 14.1 where $\sigma = 10, b = 8/3, r = 28, r_H = 24.74 \ldots$.

The bifurcation results discussed here can all be obtained with the methods of chapter 13.

5. Boundedness of the solutions.

One can find bounded, invariant sets in which the solutions are contained from some time on. We shall follow the construction of Sparrow (1982) in considering as in chapter 8 a Lyapunov function:

$$V(x, y, z) = rx^2 + \sigma y^2 + \sigma(z - 2r)^2.$$

For the orbital derivative with respect to system 14.1 we find

$$L_t V = -2\sigma(rx^2 + y^2 + bz^2 - 2brz).$$

Now we have $L_t V \geq 0$ inside and on the ellipsoid $D : rx^2 + y^2 + b(z - 2r)^2 = 4br^2$. Consider the maximum value $m$ of $V(x, y, z)$ in the full ellipsoid bounded by $D$, consider also the ellipsoid $E$ with $V(x, y, z) \leq m + \varepsilon$ ($\varepsilon$ small and positive). If a phase-point $P$ is outside $E$, it is also outside $D$ so that $L_t V(P) \leq -\delta < 0$ ($\delta$ positive). It follows that the values of $V(x, y, z)$ along the orbit which starts in $P$, have to decrease; after a finite time the orbit enters $E$ and cannot leave it again.

We conclude that in the situation where $\sigma = 10, b = 8/3, r = 28$ (figure 14.1), the flow enters an ellipsoidal domain $E$ which contains three unstable equilibrium solutions. It follows from the boundedness of $E$ and the shrinking of each volume-element in the flow that an $\omega$-limit set exists in $E$ with volume zero, see also theorem 4.3. This $\omega$-limit set may contain very irregular orbits and is called a strange attractor.

## 14.2  A mapping associated with the Lorenz-equations

It is not easy to extend our knowledge of the flow, induced by the Lorenz- equations, by elementary means. It helps to study special mappings which can be associated with the flow. It is natural to construct a Poincaré- mapping $P$ by using a two-dimensional transversal $V$ to the flow.

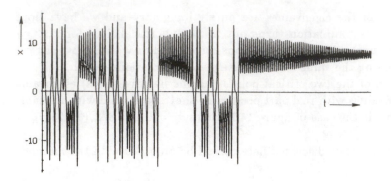

*Figure 14.2 An orbit which starts close to the strange attractor but which goes eventually to a critical point; $\sigma = 10, b = 8/3, r = 22.4$*

The points of $V$ are mapped into $V$ by the flow if the orbits return to the same region in phase-space. So to carry out the construction of $P$ correctly it is necessary to know more about phase-space, in particular about the recurrence properties of the flow. Consider for instance figure 14.2, where we have plotted the component $x(t)$ of an orbit starting in a neighbourhood of the strange attractor. First the orbit remains close to the invariant set after which it is attracted to the critical point $(\sqrt{b(r-1)}, \sqrt{b(r-1)}, r-1)$.

Another mapping was studied by Lorenz. Consider the maximal values which $z(t)$ takes subsequently for a fixed orbit. Because of the boundedness of the solutions this produces a mapping $h$ of an interval into itself.

next
maximum

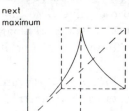

maximum in z

*Figure 14.3. Each local maximal value of $z(t)$ is followed by a new local maximum, determined by $h; \sigma = 10, b = 8/3, r = 28$*

In figure 14.3 we have plotted the local maxima horizontally which are followed by new local maxima vertically; this has been carried out for many orbits. The cusp corresponds with an orbit which goes to the origin of phase-space and which is located on the stable manifold $W_s$. To the left and to the right of the cusp, the mapping behaves in a different way which depends on the location of $W_s$. A mapping which starts in the rectangle indicated by - - -, remains there forever. If the tangent of the inclination angle of the graph is in absolute value larger than one everywhere, there exists no stable periodic orbit (the reader should check this.) It is clear that both numerical integrations and theorems on the mapping $h$ (and related mappings) can teach us a lot about the behaviour of the solutions. This also motivates us to obtain more general insight in the behaviour of mappings.

## 14.3   A mapping of $\mathbb{R}$ into $\mathbb{R}$ as a dynamical system

In the preceding section, and on many earlier occasions, we have seen that differential equations and their solutions are connected in a natural way with mappings. So it makes sense to consider more in general the dynamic behaviour of mappings. Also, in some applications like in mathematical biology, the formulation of the problem is already in the language of mappings or difference equations.

Consider as an example a population of one species; the number of individuals at time $t$ is $N_t$, $N_t \geq 0$. One unit of time later the number of individuals is $N_{t+1}$ with

$$(14.2) \qquad\qquad N_{t+1} = f(N_t)$$

where $f$ is determined by the birth and death processes in the population. We expect that $f(0) = 0$, $N_{t+1} > N_t$ if $N_t$ is small and $N_{t+1} < N_t$ if the number is large because of natural bounds on the amount of available space and food. A simple model is provided by the difference equation

$$(14.3) \qquad\qquad N_{t+1} = N_t + rN_t - \frac{r}{k}N_t^2$$

with $r$ the growth coefficient and $k$ a positive constant. Equation 14.3 is sometimes called the logistic equation. We introduce

$$x_t = rN_t/(k(1+r)) , \; a = 1 + r.$$

The difference equation 14.3 becomes

$$(14.4) \qquad\qquad x_{t+1} = ax_t(1 - x_t).$$

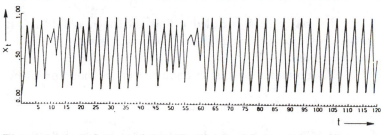

Figure 14.4. Solution of equation 14.4 with $a = 3.83$ and $x_0 = 0.1$

We choose $x \in [0, 1]$. Giving $x_0$, equation 14.4 produces the value $x_1$ for $t = 1$; substituting again produces $x_2$ for $t = 2$. We plot $x_t$ as a function of $t$, and, to obtain a better impression of the mapping, we connect the points. We carry this out with $a = 3.83$ and $x_0 = 0.1$ resulting in fairly regular behaviour of the solution for $t > 60$ (figure 14.4).

We repeat now the calculations with again $x_0 = 0.1$ but another value of $a$ : 3.99027. For $0 < t < 60$ the result doesnot look very regular (figure 14.5).

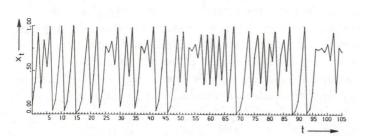

Figure 14.5. Solution of equation 14.4 with $a = 3.99027$ and $x_0 = 0.1$

Notions as "very regular" and "irregular" are too vague, so we introduce some new concepts. First we note that both autonomous differential equations and difference equations of type 14.2 can be viewed as dynamical systems.

**Definition**
$M$ is a smooth manifold; the $C^1$-mapping $\phi : R \times M \to M$ is a dynamical system for all $x \in M$ if

1. $\phi(0, x) = x$

2. $\phi(t, \phi(t_0, x)) = \phi(t + t_0, x)$
   (sometimes expressed as $\phi_t \circ \phi_{t_0} = \phi_{t+t_0}$).

In a continuous system we have $t$ and $t_0 \in R$, in a discrete system $t$ and $t_0 \in Z$.

In a continuous system the mapping $\phi$ is called a flow on the manifold $M$. It is clear that the difference equation 14.2 with $f \in C^1$ corresponds with a dynamical system; in this case we have $M \subset R$.

The phase-flow on $M \subset R^n$ , generated by the autonomous initial value problem

$$\dot{x} = f(x) , \; x(t_0) = x_0$$

is also a dynamical system. This can be seen by noting that in the definition 1 corresponds with the initial value problem formulation, 2 follows from the translation property given in lemma 2.1.

It is also possible to envisage non-autonomous $n^{th}$ order equations as a dynamical system by suspension in a $(n+1)$-dimensional autonomous system. If we have

$$\dot{x} = f(t, x)$$

we put $t = \theta$ and we derive the autonomous system

$$\dot{x} = f(\theta, x)$$
$$\dot{\theta} = 1.$$

Returning to equation 14.2 we note that $f(N_t)$ may have a fixed point $N_0 : f(N_0) = N_0$, which is called a periodic solution of the dynamical system; sometimes $N_0$ is called a periodic point of $f$. It is also possible that only after applying the mapping $k$-times, we are returning in $N_0 : f^k(N_0) = N_0$.

**Definition**
Consider the mapping $f : \mathbb{R} \to \mathbb{R}$; the point $x_0$ is called periodic point of $f$ with period $k$ if $f^k(x_0) = x_0$, $k \in \mathbb{N}$.

Also, we can formulate analogues of the definitions of stability like:

**Definition**
If $x_0$ is a fixed point of $f$, $x_0$ is asymptotically stable if there exists a neighbourhood $U$ of $x_0$ such that $\lim_{n \to \infty} f^n(x) = x_0$ for all $x \in U$. The point $x_0$ is an asymptotically stable periodic point of $f$ with period $k$ if $x_0$ is an asymptotically stable fixed point of $f^k$.

With an analogous terminology as in earlier problems we have: the domain of attraction of an asymptotically stable point $x_0$ with period $k$ is the set of points which converge to $x_0$ by iteration of the mapping $f^k$. The set $\{f^k(x_0)\}_{k=0}^{\infty}$ is called the orbit of $x_0$ .

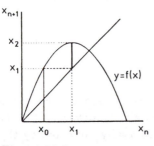

Figure 14.6

A simple graphical technique to construct subsequent mappings and to study stability runs as follows.

Consider, as in figure 14.6, the graph of $f(x)$. A vertical segment, drawn from $x_0$, intersects the graph with corresponding ordinate $x_1 = f(x_0)$. Consider a horizontal segment from this point to the line $y = x$; the point of intersection has the abscis $x_1$. Draw a vertical segment to find $x_2 = f(x_1)$ etc.

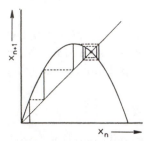

*Figure 14.7. Asymptotically stable periodic point, period 1*

It is clear that a point of intersection of the graphs of $y = f(x)$ and $y = x$ corresponds with a fixed point of $f$. In the case of figure 14.6 the fixed point is asymptotically stable. We can study this behaviour by continuing the construction as in figure 14.7. If $x_0$ is a fixed point of $f$ in 14.2 one can prove easily (the reader should check this): $\mid f'(x_0) \mid < 1$ implies that $x_0$ is asymptotically stable, from $\mid f'(x_0) \mid > 1$ it follows that $x_0$ is unstable.

In figure 14.8 we have indicated periodic points with period 2 and with period 4.

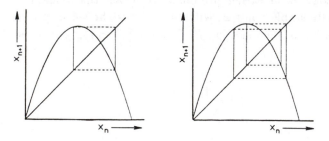

*Figure 14.8. Periodic points with period 2 and 4*

In characterising a dynamical system of the form 14.2 we discern various types among the orbits and points: periodic points and points which converge to a periodic point; there is also irregular behaviour of orbits, generally indicated by the term "chaos", but it is not easy to define this. Physicists, when studying dynamical systems, use the term chaos loosely to indicate that the system shows sensitive dependence on initial conditions and divergence of trajectories. We have met such behaviour when studying the Lorenz-equations. Mathematicians refrain from giving a general definition although in the case of one-dimensional mappings (14.2) we can be more precise:

**Definition**
A point $x_0$ is called aperiodic point of the mapping $f$ in 14.2 if the orbit of $x_0$ is bounded and if no $k \in \mathbb{N}$ exists such that $\lim_{n \to \infty} f^{nk}(x_0)$ exists. In this case the mapping $f$ is called chaotic.

If $x_0$ is aperiodic it cannot converge to a periodic point. We have seen irregular behaviour of the quadratic mapping in figure 14.5 for a particular orbit and one may ask whether the dynamical system 14.4 contains aperiodic points and, if so, where they are located. Put in a different way: consider all asymptotically stable periodic points of 14.4 with $x \in [0,1]$ and the union $A$ of all attraction domains of these points; what does the complement of $A$ in $[0,1]$ look like and what is the Lebesgue measure of the complement?

## 14.4  Results for the quadratic mapping

In this section we list a number of interesting results without being complete. For more results see Nusse (1984) and Devaney (1986) which also contain extensive surveys of the literature. We are considering again 14.4 with

$$x_{t+1} = ax_t(1 - x_t) , \; x \in [0,1].$$

The maximum of the function $ax(1-x)$ is $a/4$. So we put the requirement $0 \le a \le 4$ to ensure that 14.4 defines repeatedly a mapping of $[0,1]$ into itself. As an illustration we depict in figures 14.9 and 14.10 the function $f(x) = ax(1 - x)$ when applied $k$-times, $k = 1, \ldots, 10$.

Intersections with the graph of $y = x$ denote periodic points with period $k$. Consider first the case $a = 3.6$ in figure 14.9. For each value of $k$ one finds one or more periodic points. However, there doesnot exist for each value of $k$ a periodic point with smallest period $k$; consider for instance $k = 3$. One should note that a fixed point of $f$ produces, apart from a periodic point with period 1, periodic points with period $k$ for each $k \in \mathbb{N}$; periodic points with period 2, corresponding with fixed points of $f^2$, produce periodic points with period $2k$ for each $k \in \mathbb{N}$. Consider now the case $a = 3.99027$ in figure 14.10. The number of periodic points increases with $k$ in an explosive way. The occurrence of period 3 is very important. In 1964 Sharkovsky studied mappings of $\mathbb{R}$ into itself and he reached the following conclusion: $F$ is a continuous mapping of $\mathbb{R}$ into itself with periodic point, period 3; then for each $n \in \mathbb{N}$ there exist periodic points with smallest period $n \in \mathbb{N}$. For

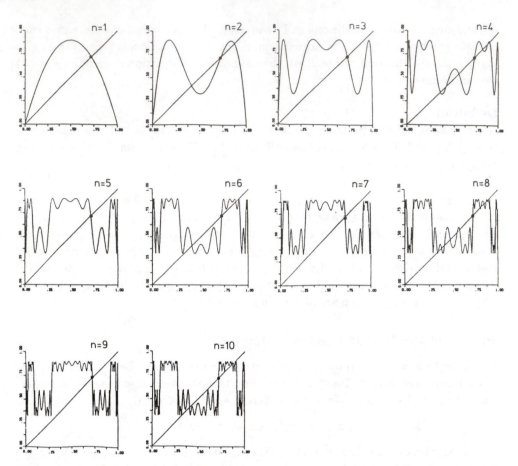

*Figure 14.9.* $f^k, k = 1, \ldots, 10$ *for* $f(x) = 3.6x(1 - x)$; *periodic points with period* $k$ *are found on the line* $y = x$. *The fixed points of* $f$ *are indicated by* •

details and a proof see Devaney (1986).

In 1975 Li and Yorke obtained a related result in a famous paper with the title "period three implies chaos". One of their results is : consider a continuous mapping $F$ of a segment into itself with periodic point, period 3; then there exist periodic points with period $n$ for each $n \in \mathbb{N}$ and there exists a noncountable set of aperiodic points in the segment.

In 1983 Nusse showed that, if the quadratic map $f$ has an asymptotically stable point $p$ and an infinite number of different periodic points, the set of points which donot converge to a stable periodic point, has Lebesgue-measure zero. In other words, a point which is randomly chosen will not be aperiodic.

Finally we call attention to the phenomenon of period-doubling and the Feigenbaum-number. The mapping $x \to ax(1 - x)$ with $x \in [0, 1]$ is now considered as a bifurcation problem with parameter $a \in [0, 4]$. We find the following bifurcations:

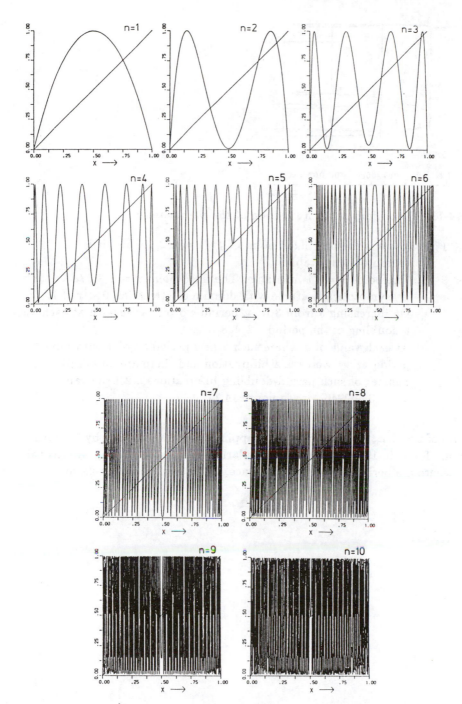

Figure 14.10. $f^k, k = 1, \ldots, 10$ for $f(x) = ax(1 - x)$ with $a = 3.99027$; periodic points with period $k$ are found on the line $y = x$

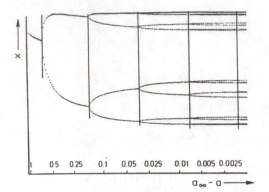

0.5   0.25    0.1    0.05  0.025    0.01  0.005 0.0025

$a_\infty - a \longrightarrow$

*Figure 14.11. The $\omega$-limit set as a function of the parameter $a$*

$0 < a \leq 1$ : $x = 0$ is the only fixed point of $f^k, k = 1, 2, \ldots$ and it is
asymptotically stable.

$1 < a < 3$ : The point $x = 0$ is unstable. The fixed point $x = 1 - 1/a$
is asymptotically stable with domain of attraction $0 < x < 1$.

$a > 3$ : On increasing $a$ from 3 to 4, periodic orbits arise with at each step
a doubling of the period : $2, 4, 8, \ldots$.
At each value of $a$ where such a new periodic point with double
period arises we have a bifurcation and there are an infinite
number of such period-doubling bifurcations until the value
$a_\infty = 3.569946\ldots$; see figure 14.11.

The values of $a$ with $a_\infty < a < 4$ produce mappings which we indicate by "chaotic".
Feigenbaum found in 1975 a remarkable regularity. If $a_1, a_2, a_3, \ldots, a_\infty$ are the values of $a$ where the period-doubling takes place, we have the simple relation

$$a_n = a_\infty - cq^{-n}$$

with $c = 2.6327\ldots$ and $q = 4.669202\ldots$ (the Feigenbaum-number). This looks like a rather technical result, but it turns out that the constant $q$ has a "universal" character in the sense that the number $q$ arises in more dynamical systems where period-doubling takes place.

# 15 Hamiltonian systems

## 15.1 Summary of results obtained earlier

In section 2.4 we were introduced to Hamiltonian systems. If $H$ is a $C^2$ function of the $2n$ variables $p_i, q_i, i = 1, \ldots, n$, $H : \mathbf{R}^{2n} \to \mathbf{R}$, then the equations of Hamilton are

$$(15.1) \qquad \dot{p}_i = -\frac{\partial H}{\partial q_i} \, , \, \dot{q}_i = \frac{\partial H}{\partial p_i}, \, i = 1, \ldots, n.$$

Now we have for the orbital derivative $L_t H = 0$, so $H(p, q)$ is a first integral of the equations 15.1. We have seen a number of examples where $n = 1$; in this case the integral $H(p, q) = $ constant describes the orbits in the phase-plane completely.
In section 8.3 we studied the stability of equilibrium solutions of the equations of Hamilton. Theorem 8.4 contains the result, that if the Hamilton function $H(p, q) - H(0, 0)$ is sign definite in a neighbourhood of the critical point $(0, 0)$, the equilibrium solution $(p, q) = (0, 0)$ is stable in the sense of Lyapunov.
Attraction by an equilibrium solution is not possible as the flow in phase- space is volume-preserving (theorem 2.1 of Liouville). This has also as a consequence that in Hamiltonian systems with one degree of freedom $(n = 1)$, the non-degenerate critical points can only be saddles or centre points. Of course, the Liouville theorem doesnot exclude the possibility of attraction in a lower-dimensional subman- ifold of $\mathbf{R}^{2n}$. For $n = 1$ this occurs at a saddle point in the corresponding stable manifold.
As an introduction to systems with more than one degree of freedom $(n > 1)$ we discuss the linear harmonic oscillator with two degrees of freedom. The Hamilton function is in our example

$$(15.2) \qquad H = \frac{1}{2}\omega_1(p_1^2 + q_1^2) + \frac{1}{2}\omega_2(p_2^2 + q_2^2).$$

We can formulate the equations of Hamilton; they lead to

$$(15.3) \qquad \ddot{q}_i + \omega_i^2 q_i = 0, \ i = 1, 2.$$

We assume now that the frequences $\omega_1$ and $\omega_2$ are positive. This Hamiltonian sys- tem generates a system with harmonic oscillations in two independent directions. Phase-space is four-dimensional and the 1-parameter family of invariant manifolds

$$(15.4) \qquad \frac{1}{2}\omega_1(p_1^2 + q_1^2) + \frac{1}{2}\omega_2(p_2^2 + q_2^2) = E_0$$

with $E_0$ a positive parameter, represents a family of ellipsoid. The value of $E_0$ is called the energy and the family 15.4 the energy-manifolds or energy-surfaces. Not only $H$ is a first integral, but in this system also

$$\tau_1 = \frac{1}{2}(p_1^2 + q_1^2) \text{ and } \tau_2 = \frac{1}{2}(p_2^2 + q_2^2).$$

Altogether there are of course only two functionally independent integrals as

$$E_0 = \omega_1 \tau_1 + \omega_2 \tau_2.$$

This existence of two independent integrals, corresponding with two families of invariant manifolds, enables us to envisage clearly the structure of phase-space. For instance fix $E_0$; the orbits are restricted now to a certain energy-surface which encloses the origin of phase-space. If $\omega_1/\omega_2 \in Q$, all solutions are periodic. If $\omega_1/\omega_2 \notin Q$, the energy-surface contains two periodic solutions, the normal modes corresponding with $\tau_1 = 0, \omega_2\tau_2 = E_0$ and with $\omega_1\tau_1 = E_0, \tau_2 = 0$.

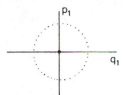

*Figure 15.1*

We shall construct a Poincaré-mapping $P$ for the flow on the energy-manifold. This manifold is an ellipsoid in $\mathbf{R}^4$ and so three-dimensional. A transversal to the flow is two-dimensional; we choose for instance the $p_1, q_1$-plane at $q_2 = 0$. The value for $p_2$ for each point of the transversal plane follows from 15.4. In the transversal $(0,0)$ corresponds with the periodic solution $\tau_1 = 0, \omega_2\tau_2 = E_0$; $(0,0)$ is a fixed point of the mapping $P$. Starting outside $(0,0)$, repeated Poincaré-mapping produces points which are located on a closed curve around $(0,0)$, see figure 15.1. For in the transversal plane we have

$$\frac{1}{2}\omega_1(p_1^2 + q_1^2) = E_0 - \omega_2\tau_2.$$

If $\omega_1/\omega_2 \notin Q$, then the starting point will never be reached again, regardless of the number of times the Poincaré-mapping is applied. If the starting point had been reached again, we would have found a periodic solution.

We repeat this process for various starting points to find a family of closed curves around $(0,0)$ in the transversal plane, see figure 15.2. The given $p_1, q_1$-plane is

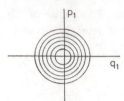

*Figure 15.2*

transversal to the periodic solution in the $p_2, q_2$-direction (normal mode) and the solutions are found on tori around this periodic solution. The existence of the second integral $\tau_2 = $ constant produces a foliation of the energy-manifold in invariant tori.

Of course we can follow the same reasoning if we choose a $p_2, q_2$- transversal plane. This plane is transversal to the periodic solution (normal mode) in the $p_1, q_1$-direction. The two normal modes which are found on the energy-surface are clearly linked as indicated in figure 15.3.

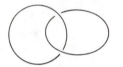

*Figure 15.3*

The foliation in invariant tori around one of the normal modes is at the same time a foliation in tori around the other normal mode. Topologically this is possible because the energy-manifold is three-dimensional.

## 15.2   A nonlinear example with two degrees of freedom

We shall consider now an example where the equations of Hamilton are nonlinear. We shall try to characterise the phase-flow by normalisation techniques. Consider the Hamiltonian

(15.5) $$H = \frac{1}{2}(p_1^2 + 4q_1^2) + \frac{1}{2}(p_2^2 + q_2^2) - q_1 q_2^2.$$

The equations of Hamilton lead to the system

(15.6) $$\begin{aligned} \ddot{q}_1 + 4q_1 &= q_2^2 \\ \ddot{q}_2 + q_2 &= 2q_1 q_2 \end{aligned}$$

It follows from theorem 8.4 that the origin of phase-space is a stable critical point. We shall consider the flow in a neighbourhood of the origin and so it makes sense to rescale

$$q_1 = \varepsilon \bar{q}_1 \ , \ q_2 = \varepsilon \bar{q}_2$$

with $\varepsilon$ a small positive parameter, $\bar{q}_1$ and $\bar{q}_2$ $O(1)$ quantities with respect to $\varepsilon$. Introducing this rescaling in system 15.6, dividing by $\varepsilon$ and omitting the bars yields the system

(15.7)
$$\ddot{q}_1 + 4q_1 = \varepsilon q_2^2$$
$$\ddot{q}_2 + q_2 = \varepsilon 2q_1 q_2.$$

Note that we can find one family of periodic solutions immediately: the family of normal modes in the $p_1, q_1$-direction. For putting $q_2 = 0$ produces $\ddot{q}_1 + 4q_1 = 0$ for the first equation and an identity for the second one.

We have some freedom in applying various related normalisation techniques. We shall use at this point the averaging method of chapter 11, we shall discuss the various techniques at the end of this section. Transformation 11.13 becomes

(15.8)
$$q_1 = r_1 \cos(2t + \phi_1) \qquad q_2 = r_2 \cos(t + \phi_2)$$
$$\dot{q}_1 = -2r_1 \sin(2t + \phi_1) \quad \dot{q}_2 = -r_2 \sin(t + \phi_2).$$

and the standard form of Lagrange

$$\dot{r}_1 = -\tfrac{1}{2}\varepsilon \sin(2t + \phi_1) r_2^2 \cos^2(t + \phi_2)$$
$$\dot{\phi}_1 = -\varepsilon \tfrac{r_2^2}{2r_1} \cos(2t + \phi_1) \cos^2(t + \phi_2)$$
$$\dot{r}_2 = -\varepsilon 2r_1 r_2 \sin(t + \phi_2) \cos(t + \phi_2) \cos(2t + \phi_1)$$
$$\dot{\phi}_2 = -\varepsilon 2r_1 \cos(2t + \phi_1) \cos^2(t + \phi_2).$$

Averaging over $t$ produces the system

(15.9)
$$\dot{\rho}_1 = -\varepsilon \tfrac{1}{8}\rho_2^2 \sin(\psi_1 - 2\psi_2)$$
$$\dot{\psi}_1 = -\varepsilon \tfrac{1}{8}\tfrac{\rho_2^2}{\rho_1} \cos(\psi_1 - 2\psi_2)$$
$$\dot{\rho}_2 = \varepsilon \tfrac{1}{2}\rho_1 \rho_2 \sin(\psi_1 - 2\psi_2)$$
$$\dot{\psi}_2 = -\varepsilon \tfrac{1}{2}\rho_1 \cos(\psi_1 - 2\psi_2).$$

Applying to $\rho_i, \psi_i$ $(i = 1, 2)$ the same initial values as to $r_i, \phi_i$ $(i = 1, 2)$, we obtain from system 15.9 an $O(\varepsilon)$ approximation, valid on the time-scale $1/\varepsilon$.

On the righthand side of system 15.9 the angles are only there in the combination $\psi_1 - 2\psi_2$. This enables us to lower the dimension of the system with one. The original system 15.7 has the energy-integral $(H = \text{constant})$, the approximating system 15.9 has two integrals:

(15.10)
$$4\rho_1^2 + \rho_2^2 = 2E_0$$

(15.11)
$$\frac{1}{2}\rho_1 \rho_2^2 \cos(\psi_1 - 2\psi_2) = I$$

with $E_0$ and $I$ constants. The first integral (15.10) is the part of the (exact) energy-integral to $O(\varepsilon)$;

15.11 is an integral for the approximate phase-flow and we donot know whether 15.11 approximates an existing integral of system 15.7. Later we shall see that system 15.7 has *no* second integral, but the phase-flow behaves near the origin as if a second integral exists corresponding with a family of invariant manifolds.

A simple type of periodic solution can be obtained when looking in system 15.9 for solutions with constant amplitudes $\rho_1$ and $\rho_2$. They exist if we have for $t \in \mathbb{R}$

$$\psi_1 - 2\psi_2 = 0 \text{ (in-phase) or } \psi_1 - 2\psi_2 = \pi \text{ (out-phase)}.$$

From the second and fourth equation of 15.9 we find

(15.12) $$\frac{d}{dt}(\psi_1 - 2\psi_2) = -\varepsilon(\frac{1}{8}\frac{\rho_2^2}{\rho_1} - \rho_1)\cos(\psi_1 - 2\psi_2).$$

So we have in-phase and out-phase periodic solutions of constant amplitude if

(15.13) $$\frac{1}{8}\frac{\rho_2^2}{\rho_1} - \rho_1 = 0.$$

From condition 15.13 and the approximate energy-integral 15.10 we find

$$\rho_1 = \sqrt{E_0/6} \, , \; \rho_2 = \sqrt{4E_0/3}.$$

From these values of $\rho_1, \rho_2$ and $\psi_1 - 2\psi_2$ system 15.9 becomes

$$\dot{\rho}_1 = 0, \dot{\psi}_1 = -\varepsilon\sqrt{E_0/6}(\pm 1), \dot{\rho}_2 = 0, \dot{\psi}_2 = -\varepsilon\frac{1}{2}\sqrt{E_0/6}(\pm 1).$$

The in-phase periodic solution is

$$\begin{aligned} q_1(t) &= \sqrt{E_0/6}\cos(2t + \phi_1(0) - \varepsilon\sqrt{E_0/6}t) + O(\varepsilon) \\ q_2(t) &= \sqrt{4E_0/3}\cos(t + \tfrac{1}{2}\phi_1(0) - \varepsilon\tfrac{1}{2}\sqrt{E_0/6}t) + O(\varepsilon) \end{aligned}$$

For the out-phase periodic solution we find

$$\begin{aligned} q_1(t) &= \sqrt{E_0/6}\cos(2t + \phi_1(0) + \varepsilon\sqrt{E_0/6}t) + O(\varepsilon) \\ q_2(t) &= \sqrt{4E_0/3}\cos(t + \tfrac{1}{2}\phi_1(0) - \tfrac{\pi}{2} + \varepsilon\tfrac{1}{2}\sqrt{E_0/6}t) + O(\varepsilon) \end{aligned}$$

The approximations are valid on the time-scale $1/\varepsilon$. In these expressions $E_0$, the value of the energy, is a free parameter. So, apart from the family of normal modes in the $p_1, q_1$-direction which we found earlier, we have discovered two new families of periodic solutions. We donot find a continuation of the normal modes which in the linearised system ($\varepsilon = 0$) exists in the $p_2, q_2$-direction.

For a given value of $E_0$ we have three periodic solutions on the energy- manifold; the behaviour of the phase-flow on the energy-manifold around these periodic solutions can be deduced from the second integral 15.11. We rearrange

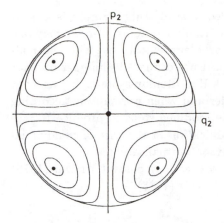

*Figure 15.4. Poincaré-mapping for system 15.6*

$$\rho_1\rho_2^2\cos(\psi_1 - 2\psi_2) = \rho_1\rho_2^2\cos(2t + \psi_1 - 2t - 2\psi_2) =$$
$$= \rho_1\rho_2^2[\cos(2t + \psi_1)\cos(2t + 2\psi_2) + \sin(2t + \psi_1)\sin(2t + 2\psi_2)].$$

As $\dot{q}_1 = p_1, \ldots q_2 = p_2$ we find with 15.8 and 15.11

(15.14) $$I - \frac{1}{2}(q_1q_2^2 - q_1p_2^2 + p_1p_2q_2) = O(\varepsilon)$$

on the time-scale $1/\varepsilon$. We use again Poincaré-mapping and we choose a $p_2, q_2$-transversal to the flow for fixed $E_0$; choose for instance $q_1$ and eliminate $p_1$ with the energy-integral. The (approximate) integral 15.14 produces the Poincaré-mapping of figure 15.4; the map contains one saddle and four centre points. The saddle in the origin is the fixed point corresponding with the normal mode in the $p_1, q_1$-direction; this periodic solution is clearly unstable. The centre points in the first and third quadrants and in the second and fourth quadrants belong to each other, they correspond with the two other periodic solutions which we found. These solutions are recognized as fixed points of $P^2$ and they are both stable. Around the stable periodic solutions the phase-flow moves on invariant tori of which the intersections with the transversal plane have been drawn.

The analysis which led to the Poincaré-mapping of figure 15.4 has been based on an approximation of the Hamiltonian vector field. It is clear that by computing higher order approximations we can obtain more precise results. However, the question is whether we have found all qualitatively different phenomena which play a part in this example. It turns out that this is not the case. The periodic solutions which we have found do exist for system 15.7 and the stable ones have families of invariant tori around them. However, between the tori the orbits show irregular behaviour, we shall say more about this in section 15.5.

Normalisation by averaging, which we have used to study this example, is a technique which has produced many results for Hamiltonian systems. For a survey one may consult Verhulst (1983). One of the first problems which arise when normalising Hamiltonian systems, is that simple-minded use of these techniques doesnot quarantee that the normalised system is again a Hamiltonian system. It turns out that both when using averaging to higher order and when normalising according to Poincaré and Dulac, the calculations can be modified so as to produce Hamiltonian systems after normalisation. This modification in the case of polynomial transformations is often called Birkhoff-Gustavson normalisation.

Apart from the survey mentioned above, the reader may consult Arnold (1976) appendix 7, Sanders and Verhulst (1985) and Arnold (1988).

## 15.3 The phenomenon of recurrence

In the Liouville theorem (2.1) we have seen that the flow induced by a time- independent Hamiltonian is volume-preserving. In the problems which we studied in sections 15.1 and 15.2, the flow takes place on bounded energy- manifolds. In that case nearly all orbits will return after some time to a neighbourhood of the starting point. This is a consequence of the recurrence theorem:

**Theorem 15.1 (Poincaré)**
Consider a bounded domain $D \subset \mathbb{R}^n$, $g$ is a bijective, continuous, volume-preserving mapping of $D$ into itself. Then each neighbourhood $U$ of each point in $D$ contains a point $x$ which returns to $U$ after repeated application of the mapping ($g^n x \in U$ for some $n \in \mathbb{N}$).

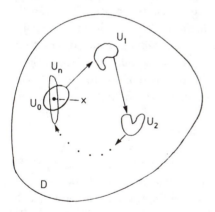

*Figure 15.5*

**Proof**
Consider a set $U_0 \in D$ with positive volume. We apply the map $g$ to the points of

$U_0 : gU_0 = U_1$, $gU_1 = U_2$, etc. The sets $U_0, U_1, U_2$, etc. have all the same volume. As the volume of $D$ is finite, the sets $U_0, U_1, U_2$, etc. cannot be all disjunct. Suppose that $g^i U_0 \cap g^k U_0 \neq \phi \ (k > i)$.

We note that $g$ is a dynamical system in the sense of section 14.3, so we have

$$U_0 \cap g^{k-i} U_0 \neq \phi.$$

Put $k - i = n$; so in $U_0$ we can find points which return into $U_0$ by the mapping $g^n$.                                                                            □

### Remarks

We have followed the formulation and the proof by Poincaré. In modern formulations of the theorem one uses instead of "volume" the concept (Lebesgue-) "measure" and for "volume-preserving mapping" the term "measure-preserving mapping."

A dynamical system which satisfies the requirements of the recurrence theorem is called "stable in the sense of Poisson". This terminology derives also from Poincaré who was considering in this respect Poisson's (unproved) propositions on the stability of the solar system.

### Example 15.1

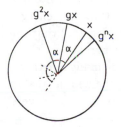

*Figure 15.6*

For $D$ we choose a circle $(S^1)$ and for $g$ : rotation over an angle $\alpha > 0$. The requirements of the recurrence theorem have been satisfied. There are two possibilities for recurrent dynamics of the system.

a. $\alpha/2\pi = m/n$ with $m, n \in \mathbb{N}$ and relative prime. Then $g^n x = x$ and all orbits of the dynamical system are periodic.

b. $\alpha/2\pi \notin Q$. No periodic orbits exist in this dynamical system. However, according to theorem 15.1, for each $\delta > 0$ there exists a number $n \in \mathbb{N}$ such that $| g^n x - x | < \delta$.

### Example 15.2

An example of a Hamiltonian system with one degree of freedom is the equation in example 8.2

$$\ddot{x} + f(x) = 0.$$

In this example we showed that on finding after linearisation a centre point, the corresponding equilibrium solution is stable and the orbits in a neighbourhood of the centre point in the phase-plane are closed. Theorem 15.1 applies but the case is rather trivial as all the solutions in a neighbourhood of the equilibrium solution are periodic. Note that the simple example 15.1 shows already more complicated dynamical behaviour.

**Example 15.3**
Linear Hamiltonian system with two degrees of freedom.
The linear equations discussed in section 15.1 are

$$\ddot{q}_1 + \omega_1^2 q_1 = 0 \, , \ \ddot{q}_2 + \omega_2^2 q_2 = 0.$$

Theorem 15.1 applies. If $\omega_1/\omega_2 \in Q$, we have again the trivial situation of periodic orbits only. If $\omega_1/\omega_2 \notin Q$, the orbits are moving quasi-periodically over the tori around the two normal modes for each value of the energy. Dynamically the situation looks like example 15.1 in the irrational case.

The remarkable fact is of course that theorem 15.1 applies to any Hamiltonian system of the form 15.1 with $n$ degrees of freedom provided that the energy-manifold is bounded. In particular we have in the problem of section 15.2 that the phase-flow in a neighbourhood of the origin of phase-space is recurrent.
For some problems in mechanics, for instance for systems with many particles, this phenomenon of recurrence seems to be a curious result. One should realise however that the recurrence time for a system with many particles can be very long.
Finally we remark that application of the recurrence theorem to systems with three or more particles bound by gravity (think of models for the solar system) is difficult as the energy-manifold is not bounded. One can find some invariant subsets where the recurrence theorem applies, see Arnold (1988) section 2.6.

## 15.4 Periodic solutions

The demonstration of the existence of families of periodic solutions of Hamiltonian systems is carried out by using fixed-point theorems, variational methods and analytic-numerical techniques; we shall only discuss a basic result here. The theory of the stability of periodic solutions is more open and contains difficult mathematical questions.
Most theorems have been obtained for systems in a neighbourhood of a stable

critical point. In many cases the Hamilton function can be written as

$$(15.15) \qquad H = \sum_{i=1}^{n} \frac{1}{2}\omega_i(p_i^2 + q_i^2) + \ldots, \quad \omega_i > 0$$

where the dots represent cubic and higher-order terms. We shall indicate solutions by short-periodic if the period is of the order of magnitude $2\pi/\omega_i$; the indication "short" has nothing to do with a small parameter.

A basic result is the theorem of Weinstein (1973) which tells us that a Hamiltonian system with $n$ degrees of freedom in a neighbourhood of a stable critical point (in the case of 15.15) there exist at least $n$ short-periodic solutions for each value of the energy. In this respect we remark that in the linear problem of section 15.1 we have found for each value of the energy two or an infinite number of periodic solutions, in the nonlinear problem of section 15.2 we have found three periodic solutions for each small value of the energy.

An interesting question is whether the periodic solutions corresponding with the normal modes of the linearised problem ($\varepsilon = 0$ in section 15.2) can be continued to periodic solutions of the nonlinear problem. In the case of the system in section 15.2 one normal mode can be continued, for the other normal mode this seems not to be the case. Lyapunov demonstrated for the Hamiltonian 15.15 that if the frequencies $\omega_i, i = 1, \ldots, n$ are independent over $\mathbf{Z}$, the normal modes of the linear problem can be continued to periodic solutions of the nonlinear problem. This result also holds if some of the $\omega_i$ are negative. The periodic solutions, according to this theorem of Lyapunov, are located in two-dimensional manifolds which in the origin of phase-space are tangent to the $p_i, q_i$-coordinate planes ($i = 1, \ldots, n$).

Lyapunov's theorem is concerned with the non-resonant case; the Weinstein theorem admits resonances but it has to drop the idea of continuation and the energy-manifold has to be bounded.

## 15.5  Invariant tori and chaos

The modern theory of Hamiltonian systems has been developed under the influence of two important stimuli: the numerical calculations of Hénon and Heiles, Ollongren, Contopoulos and later many others and on the other hand the theorem of Kolmogorov, Arnold and Moser (KAM).

First we describe the computations of Hénon and Heiles which have been very influential because of the questions they raised. Consider the Hamiltonian

$$(15.16) \qquad H = \frac{1}{2}(p_1^2 + q_1^2) + \frac{1}{2}(p_2^2 + q_2^2) + q_1^2 q_2 - \frac{1}{3}q_2^3.$$

The system has two degrees of freedom and the equations of Hamilton lead to the system

$$(15.17) \qquad \begin{aligned} \ddot{q}_1 + q_1 &= -2q_1 q_2 \\ \ddot{q}_2 + q_2 &= -q_1^2 + q_2^2. \end{aligned}$$

System 15.17 is integrated numerically for various values of the energy $E_0$ and many initial values. The results of these computations are used to construct a Poincaré-mapping of the $p_2, q_2$-transversal plane at $q_1 = 0$; $p_1$ follows from the energy integral where we choose the positive root.

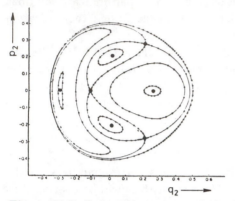

*Figure 15.7. Poincaré transversal plane in system 15.17 for $E_0 = 1/12$; after Hénon and Heiles, Astron. J. 69, p.75 (1964)*

On choosing $E_0 = 1/12$, the transversal plane shows a regular pattern with 7 fixed points and closed curves around 4 of them. The fixed points correspond with periodic solutions, the closed curves correspond with tori which are lying around the stable periodic solutions. The boundary is not transversal to the flow, it corresponds with a periodic solution ($q_1(t) = 0$ for $t \in \mathbb{R}$ produces an identity for the first equation in 15.17 and, for sufficiently small values of $E_0$, a periodic solution in the second equation); so there are 8 periodic solutions for this value of the energy, 4 of them are stable.

In the discussions of Poincaré-mappings some technical terms frequently occur which are used to describe special orbits. If one starts in the transversal plane of figure 15.7 in a point of a curve connecting two different saddle points, the corresponding orbit is called *heteroclinic*. As we know, the saddle points correspond with periodic solutions and a heteroclinic orbit has the property that for $t \to +\infty$ the orbit approaches one periodic solution, for $t \to -\infty$ the heteroclinic orbit approaches another periodic solution.

If the transversal plane contains a saddle with a closed loop, an orbit starting on such a loop is called *homoclinic*. It has the property that for $t \to +\infty$ the orbit approaches a periodic solution, for $t \to -\infty$ the homoclinic orbit approaches the same periodic solution.

Hénon and Heiles repeated the numerical calculations at a higher value of the energy, $E_0 = 1/8$. The consequences for the Poincaré transversal plane are remarkable. Again we find 4 fixed points which correspond with 4 stable periodic solutions around which there are tori.

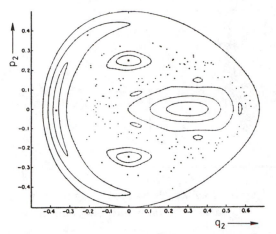

*Figure 15.8. Poincaré transversal plane in system 15.17 for $E_0 = 1/8$*

However, many tori have vanished and they have been replaced by an irregular pattern of many orbits. In figure 15.8 all points outside the closed curves correspond with just one solution which produces a new point, each time that we apply the Poincaré-mapping.

On increasing the energy still more, the domain covered by the tori decreases in size.

The general impression is that for small values of the energy, for instance $E_0 = 1/12$, the energy-manifold is foliated into invariant tori, so that apart from the energy-integral a second integral has to exist. At higher values of the energy, at a certain value of $E_0$, tori are being destroyed and irregular behaviour arises. This interpretation is not quite correct. The fact is that invariant tori exist around the stable periodic solutions but they donot cover the energy-manifold completely, even for very small values of the energy. Between the tori there exist irregular orbits, which together have a measure which tends to zero as $E_0$ tends to zero. It can be shown that these irregular orbits arise near the stable and unstable manifold of the saddles in the Poincaré-mapping i.e. near the homoclinic and heteroclinic orbits; see also Guckenheimer and Holmes (1983) chapter 4.

Precise numerical calculations by Magnenat have illustrated this phenomenon. These calculations show that at the value of the energy $E_0 = 1/12$ one finds irregular behaviour in a neighbourhood of the unstable periodic orbits, see figure 15.9. We can put this in a different way. Although the measure of the set of invariant tori tends to one as the energy $E_0$ tends to zero, we can always find initial values corresponding with orbits which are not located on invariant tori. So, a second integral apart from the energy doesnot exist; however, this statement is slightly misleading as the number of invariant tori increases sharply as $E_0$ tends to zero and, as in the problem of section 15.2, an approximate second integral exists. In the next section we shall call a system like the Hénon-Heiles problem or the exam-

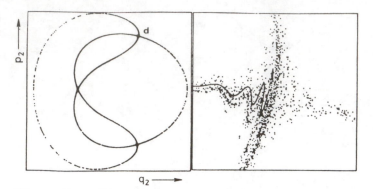

Figure 15.9. *The transversal plane of figure 15.7 at* $E_0 = 1/12$ *magnified with a factor 23 around the saddle fixed point d (after P. Magnenat, Astronomy and Astrophysics, 77, p.337, 1979)*

ple in section 15.2, nearly-integrable or asymptotically integrable for small values of the energy.

## 15.6  The KAM theorem

We shall now introduce some concepts which enable us to discuss the KAM theorem on the existence of invariant tori.

**Definition**
Consider the equations of Hamilton $\dot{p} = -\partial H/\partial q, \dot{q} = \partial H/\partial p$ (15.1) with integrals $F_1(p, q)$ and $F_2(p, q)$. The functions $F_1$ and $F_2$ are in *involution* (or *commute*) if

$$\{F_1, F_2\} = \frac{\partial F_1}{\partial q}\frac{\partial F_2}{\partial p} - \frac{\partial F_1}{\partial p}\frac{\partial F_2}{\partial q} = 0$$

($\{,\}$ are the so-called Poisson-brackets).

Note that $F$ is an integral of system 15.1 if

$$
\begin{aligned}
L_t F(p, q) &= \frac{\partial F}{\partial p}\dot{p} + \frac{\partial F}{\partial q}\dot{q} \\
&= -\frac{\partial F}{\partial p}\frac{\partial H}{\partial q} + \frac{\partial F}{\partial q}\frac{\partial H}{\partial p} = 0
\end{aligned}
$$

So $F$ is an integral if $F$ and $H$ are in involution.

**Definition**
The $n$ degrees of freedom Hamiltonian system $\dot{p} = -\partial H/\partial q, \dot{q} = \partial H/\partial p$ (15.1) is *integrable* if the system has $n$ integrals $F_1, \ldots, F_n$ which are functionally independent and in involution. In other words: the system is integrable if for $i, k = 1, \ldots, n$

$$\{F_i, H\} = 0, \quad \{F_i, F_k\} = 0,$$

with $F_1, \ldots, F_n$ functionally independent.

## Example 15.4

The quadratic Hamiltonian $H(p,q) = \sum_{i=1}^{n} \frac{1}{2}\omega_i(p_i^2+q_i^2)$ generates an integrable system with integrals $F_i = p_i^2 + q_i^2$, $i = 1,\ldots,n$.

## Example 15.5

If the Hamiltonian doesnot depend on $q$, $H = H(p)$, the system is integrable. The integrals are $F_i = p_i$, $i = 1,\ldots,n$.

A Hamiltonian system being integrable is an exceptional case but, as discussed with regards to the Hénon-Heiles problem, the statement is slightly misleading. A number of important examples in applications, for instance the Kepler problem, are integrable Hamiltonian systems and, more importantly, a number of Hamiltonian systems behave as an asymptotically integrable system when in the neighbourhood of a stable equilibrium solution. The last statement holds generally for two degrees of freedom systems and it holds for systems with more degrees of freedom in the presence of certain symmetries.

In the case of an integrable system the structure of phase-space is relatively simple. All orbits are moving on surfaces defined by the relations $F_i(p,q) = $ constant, $i = 1,\ldots,n$. In our examples until now, these were the energy-surface and the invariant tori which we found. In section 11.5 we introduced action-angle coordinates and they are useful here too. These variables, which are related to polar coordinates, have the property that when introducing them, one conserves the Hamiltonian character of the system. Such transformations are called *canonical*. In the discussion of section 11.5 we transformed $p, q \rightarrow I, \phi$ where we used the generating function $S(I,q)$. $S$ has the property that

$$(15.18) \qquad p = \frac{\partial S}{\partial q}, \phi = \frac{\partial S}{\partial I}, \; H(p,q) = H(\frac{\partial S}{\partial q},q) = H_0(I).$$

If we can find a generating function $S(I,q)$ with these properties, the Hamiltonian system 15.1 becomes

$$(15.19) \qquad \dot{I} = 0, \dot{\phi} = \omega(I) = \frac{\partial H_0}{\partial I}.$$

In general we will not be able to find such a generating function as the existence of $S$ immediately leads to integrability of the system, integrals $I,\ldots,I_n$. The transformations are especially useful if system 15.1 is *nearly-integrable* in the following sense.

Supposing that $H(p,q)$ contains a small parameter $\varepsilon$ and introduction of action -

angle coordinates produces the system

(15.20)
$$\begin{aligned} \dot{I} &= \varepsilon f(I, \phi) \\ \dot{\phi} &= \omega(I) + \varepsilon g(I, \phi). \end{aligned}$$

If $\varepsilon = 0$ the system is integrable; if $0 < \varepsilon \ll 1$ the system is called nearly-integrable. System 15.20 can be studied by suitable perturbation- and normalisation methods.

Suppose now that the Hamiltonian of system 15.20 is

(15.21)
$$H_0(I) + \varepsilon H_1(I, \phi).$$

The following theorem guarantees that many of the invariant tori which exist in the integrable case $\varepsilon = 0$, also exist for $\varepsilon > 0$, be it somewhat deformed by the perturbation.

**Theorem 15.2 (KAM)**

Consider the system of equations 15.20 induced by the analytic Hamiltonian 15.21. If $H_0$ is non-degenerate i.e.

$$\det \left( \frac{\partial^2 H_0}{\partial I^2} \right) \neq 0,$$

then most of the invariant tori which exist for the unperturbed system ($\varepsilon = 0$) will, slightly deformed, also exist for $\varepsilon > 0$ and sufficiently small; moreover the (Lebesque-) measure of the complement of the set of tori tends to zero as $\varepsilon$ tends to zero.

The basis of the proof is the construction of successive changes of coordinates which remove in the Hamiltonian the terms depending on the angles. This process turns out to be quadratically convergent in $\varepsilon$ if one excludes certain "resonant sets" in phase-space. The analyticity condition on $H$ has been reduced by Moser to smoothness of finite order. The reader should consult Arnold (1976) appendix 8 and Arnold (1988) chapter 5 for a more detailed discussion and references.

For systems with two degrees of freedom as we have seen in section 15.2, we may visualise the Poincaré-mapping as in figure 15.10.

Around a stable periodic solution there are tori on which the orbits move quasi-periodically. In between the tori one has irregular behaviour which vanishes as $\varepsilon$ tends to zero, but which increases in importance if $\varepsilon$ increases. The last phenomenon has been illustrated by the computations of Hénon and Heiles; see again the figures 15.7 and 15.8.

Finally some remarks about the stability of periodic solutions and its relation with the existence of invariant tori. When linearising around periodic solutions

*Figure 15.10*

of time-independent $n$ degrees of freedom Hamiltonian systems we find at least two of the characteristic exponents to be zero. One of these zeros arises because of the property of periodic solutions of autonomous systems discussed in section 7.3; the other zero arises as periodic solutions of Hamiltonian systems in general occur in one-parameter families, parameterised for instance by $E_0$. These two zeros complicate the stability analysis but, as we showed in section 15.2, this degeneracy can be removed by two reduction steps. First we fix the energy and in general we will then have an isolated periodic solution on the energy manifold. Then we choose a Poincaré transversal plane to find periodic solutions as fixed points of the corresponding Poincaré-mapping.

*Figure 15.11. Eigenvalues of fixed points of Poincaré-mapping in three degrees of freedom Hamiltonian systems; the location is symmetric with respect to the axes in the complex plane*

In the case of two degrees of freedom, phase-space has dimension four, the transversal plane has dimension two. Because of the volume-preserving character of the flow a non-degenerate fixed point can only be a saddle or a centre point in the linear approximation. Nonlinear perturbations cannot change this picture as the case of two complex eigenvalues is not allowed. Geometrically this can be visualised as follows. The invariant tori are described by rotation over the two angles, keeping the two actions fixed, so the tori are two-dimensional. The tori are surrounding the stable periodic solutions and, as the energy manifold is three-dimensional, they separate the energy manifold in domains from which the solutions cannot escape. In the case of three degrees of freedom, phase-space has dimension six, the transversal has dimension four. Because of the volume-preserving character of the flow a non-degenerate fixed point of the Poincaré-mapping falls into one of four categories; see figure 15.11. Perturbation of purely imaginary eigenvalues may lead to complex eigenvalues, the stability analysis is not complete if one finds only purely imaginary eigenvalues by linearisation near the fixed point.

Geometrically one has the following picture. The invariant tori are described by rotation over the three angles, keeping the three actions fixed, so the tori are three-dimensional.

We summarize these dimensions for $2, 3$ and $n$ degrees of feedom:

| number of degrees of freedom | 2 | 3 | $n$ |
|---|---|---|---|
| dimension of phase-space | | 4 | 6 | $2n$ |
| dimension of energy manifold | | 3 | 5 | $2n - 1$ |
| dimension of invariant tori | | | 2 | 3 | $n$ |

In the case of three degrees of freedom the energy-manifold is 5-dimensional and the tori donot separate anymore the energy manifold into non-communicating invariant sets. This means that in the case of three or more degrees of freedom the orbits can "leak away" between the tori; this process has a very long time-scale and is called "Arnold-diffusion". The theory of Hamiltonian systems with three and more degrees of freedom has still many interesting open problems, both of a quantitative and a qualitative nature.

## 15.7 Exercises

15-1. Consider the equations of motion induced by a time-independent $n$ degrees of freedom Hamiltonian. We are interested in the possible eigenvalues of critical points and periodic solutions, leaving out degeneracies.

  a. Take $n = 2$; what are the possible eigenvalues of the critical points.

  b. The same question for the periodic solutions if $n = 2$.

  c. Take $n = 3$; what are the possible eigenvalues of the periodic solutions.

15-2. The 2 degrees of freedom system derived from the Hamiltonian

$$H = \frac{1}{2}\omega_1(p_1^2 + q_1^2) + \frac{1}{2}\omega_2(p_2^2 + q_2^2) + \varepsilon H_3 + \varepsilon^2 \dots$$

is studied near the origin of phase-space. The frequencies $\omega_1$ and $\omega_2$ are positive, $H_3$ is homogeneous of degree 3, the dots represent terms of degree 4 and higher. We simplify the system by normalisation.

a. Show that we have resonance.

b. Resonance which results in a normal form consisting of 2 decoupled one-dimensional oscillators neednot worry us. If there is coupling in the normal form via the cubic terms of the Hamiltonian we call this first-order resonance. For which values of $w_1$ and $w_2$ does this occur?

c. If the resonant coupling starts in the quartic terms we have a second-order resonance. For which values of $w_1$ and $w_2$ does this occur?

d. What is the time-scale of validity of the approximations derived from the normal form in both cases.

15-3. We generalise the example of section 15.2 in two ways: we are looking at a slightly more general potential problem and we admit a small perturbation of the frequency ratio (note that an exact frequency ration $2 : 1$ cannot be realised in mechanics).

$$H = \frac{1}{2}(p_1^2 + w^2 q_1^2) + \frac{1}{2}(p_2^2 + q_2^2) - \varepsilon(\frac{1}{3}aq_1^3 + bq_1 q_2^2)$$

with $w^2 = 4(1 + c\varepsilon)$, $a, b$ and $c$ parameters.

a. Use polar coordinates (15.8) to obtain the Lagrange standard form.

b. Average the system.

c. Find the integrals of the averaged system.

d. Find the periodic solutions and the bifurcation values of $c$.

e. Determine the stability of the periodic solutions.

15-4. Consider the system

$$\ddot{q}_1 + q_1 = aq_1^2 + bq_2^2$$
$$\ddot{q}_2 + \ddot{q}_2 = 2bq_1 q_2$$

a. Show that the stationary solution $q_1 = \dot{q}_1 = q_2 = \dot{q}_2 = 0$ is stable.

b. Find families of periodic solutions which are not restricted to small values of the energy (hint: put $q_1 = \lambda q_2$).

15-5. We apply the results of exercise 15.4 to the Hénon-Heiles system 15.17 in the form of exercise 15.4: $a = 1$, $b = -1$, $q_1$ and $q_2$ have been interchanged.

$$H = \frac{1}{2}(p_1^2 + q_1^2) + \frac{1}{2}(p_2^2 + q_2^2) - \frac{1}{3}q_1^3 + q_1 q_2^2$$

a. Determine the critical points and their stability.

b. For which values of $H$ is the energy-surface compact (bounded and closed).

c. Show that periodic solutions exist in the chaotic region described in section 15.5.

d. In c we showed that three families of periodic solutions exist between two critical energy levels 0 and 1/6 where a critical point is foud. This is analogous to the one degree of freedom case. Might one conjecture that in two degrees of freedom Hamiltonian systems families of periodic solutions are forming manifolds which are connecting critical points as in this example?

15-6. Consider the two-dimensional system $\dot{x} = f(x,y), \dot{y} = g(x,y)$ which has a first integral $F(x,y) = $ constant.

a. Show that by a formal transformation of time one can put the system in Hamiltonian form.

b. Why is the transformation called formal?

c. Apply this result to the Volterra-Lotka equations from example 2.15.

d. Can we generalise this procedure to a four-dimensional autonomous system with a first integral?

# Appendix 1: The Morse lemma

In chapter 2 we have formulated lemma 2.3:
Consider the $C^\infty$-function $F : \mathbb{R}^n \to \mathbb{R}$ with non-degenerate critical point $x = 0$, index $k$. In a neighbourhood of $x = 0$ there exists a diffeomorphism $x \to y$ which transforms $F(x)$ into

$$G(y) = F(0) - y_1^2 - y_2^2 - \ldots - y_k^2 + y_{k+1}^2 + \ldots + y_n^2.$$

**Proof (Milnor, 1963)**
The function $F$ will be written as

$$
\begin{aligned}
F(x_1,\ldots,x_n) &= F(0) + \int_0^1 \frac{dF}{dt}(tx_1,\ldots,tx_n)dt \\
&= F(0) + \int_0^1 \sum_{i=1}^n \frac{\partial F}{\partial x_i}(tx_1,\ldots,tx_n)x_i\,dt.
\end{aligned}
$$

Putting $f_i(x_1,\ldots,x_n) = \int_0^1 \frac{\partial F}{\partial x_i}(tx_1,\ldots,tx_n)dt$ we have

$$F(x_1,\ldots,x_n) = F(0) + \sum_{i=1}^n x_i f_i(x_1,\ldots,x_n).$$

Because $0$ is a critical point one must have $f_i(0) = \frac{\partial F}{\partial x_i}(0) = 0$. We repeat this process once for the functions $f_i$. Then one finds

$$f_j(x_1,\ldots,x_n) = \sum_{i=1}^n x_i h_{ij}(x_1,\ldots,x_n)$$

with $h_{ij}$ a $C^\infty$-function. So we may write

$$F(x_1,\ldots,x_n) = F(0) + \sum_{i,j=1}^n x_i x_j h_{ij}(x_1,\ldots,x_n).$$

This expansion can be rearranged by putting

$$\bar{h}_{ij} = \frac{1}{2}(h_{ij} + h_{ji})$$

so that

$$F(x_1,\ldots,x_n) = F(0) + \sum_{i,j=1}^n x_i x_j \bar{h}_{ij}(x_1,\ldots,x_n)$$

with $\bar{h}_{ij} = \bar{h}_{ji}$. As the critical point is non-degenerate, the matrix $(\bar{h}_{ij}(0))$ is non-singular. So we have obtained a symmetric quadratic form for $F$ which we can diagonalise in a neighbourhood of $x = 0$. The first term is easy to obtain by transforming

$$y_1 = \mid \bar{h}_{11} \mid^{\frac{1}{2}} (x_1 + \sum_{i=2}^n x_i \bar{h}_{i1}/ \mid \bar{h}_{11} \mid)$$

Squaring this expression produces the desired result. Suppose now that we have carried on the process until $i = r - 1$ so that

$$F = F(0) \pm y_1^2 \pm \ldots \pm y_{r-1}^2 + \sum_{i,j \geq r}^{n} y_i y_j H_{ij}(y_1, \ldots, y_n).$$

The plus or minor sign is determined by the sign of $\bar{h}_{ii}$ in a neighbourhood of 0. The coefficients $\bar{h}_{ij}$ have been changed into $H_{ij}$ by the transformations but they are keeping their symmetry. Now put

$$v_i = y_i \ , \ \ i \neq r$$

$$v_r = |H_{rr}|^{1/2} \, (y_r + \sum_{i=r+1}^{n} y_i H_{ir} / |H_{rr}|).$$

After squaring this expression and substitution we have

$$F = F(0) \pm v_1^2 \pm \ldots \pm v_r^2 + \sum_{i,j=r+1}^{n} v_i v_j H'_{ij}(v_1, \ldots, v_n)$$

so that we will have the result of the lemma by induction; after determining the signs of the quadratic terms we order them according to minus or plus sign.

# Appendix 2: Linear periodic equations with a small parameter

In section 6.3 we considered linear periodic equations. We shall now study such equations in which the periodic term acts as a perturbation:

$$\dot{x} = Ax + \varepsilon B(t, \varepsilon)x$$

with $x \in \mathbf{R}^n$, $A$ and $b$ non-singular $n \times n$-matrices; $B$ is continuous in $t$, $T$-periodic and $C^2$ in $\varepsilon$; $0 \leq \varepsilon \leq \varepsilon_0$.

With these assumptions the characteristic exponents are continuous functions of $\varepsilon$ in a neighbourhood of $\varepsilon = 0$.

If there are characteristic exponents with real part zero (multipliers $\pm 1$) for certain values of the parameter $\varepsilon$ and the coefficients in $A$ and $B$, these values can be the transition cases between unstable and stable solutions. In such a transition case the system has periodic solutions. Looking for periodic solutions by using perturbation theory we are determining the boundaries of the stability domains. We shall illustrate this idea for the Mathieu-equation:

$$\ddot{x} + (a + \varepsilon \cos 2t)x = 0, \quad a > 0.$$

The equation is $\pi$-periodic. We are interested in knowing for which values of $a$ and $\varepsilon$ the trivial solution $x = 0$ is stable. If $\rho_1$ and $\rho_2$ are the multipliers, $\lambda_1$ and $\lambda_2$ the exponents, we have from theorem 6.6

$$\rho_1 \rho_2 = 1 , \quad \lambda_1 + \lambda_2 = 0.$$

The exponents are functions of $\varepsilon$, $\lambda_1 = \lambda_1(\varepsilon)$, $\lambda_2 = \lambda_2(\varepsilon)$ with clearly $\lambda_1(0) = i\sqrt{a}$, $\lambda_2(0) = -i\sqrt{a}$. As $\lambda_1 = -\lambda_2$, the exponents cannot be complex with nonzero real part. The only possibility of nonzero real parts in a neighbourhood of $\varepsilon = 0$ are clearly the cases $\sqrt{a} = m$, $m = 1, 2, \ldots$ as then the imaginary part of $e^{\lambda t}$ belongs to the term $P(t)$ in theorem 6.5 of Flocquet.

We conclude that for $a \neq m^2$, $m = 1, 2, \ldots$ we have stability for $\varepsilon$ in a neighbourhood of zero. The cases $a = m^2$ have to be considered separately. We shall determine now $a = a(\varepsilon)$ such that for $m = 1, 2, \ldots$ the Mathieu-equation has periodic solutions. The relation between $a$ and $\varepsilon$ determines the boundary between staility and instability for various values of $a$ and $\varepsilon$.

We put $a = m^2 - \varepsilon\beta$ and we use transformation 11.5 :

$$x(t) = y_1(t) \cos mt + y_2(t) \sin mt$$
$$\dot{x}(t) = -my_1(t) \sin mt + my_2(t) \cos mt.$$

Transforming $x, \dot{x} \to y_1, y_2$ yields with the Mathieu- equation

$$A2 - 1 \qquad \begin{aligned} \dot{y}_1 &= -\tfrac{\varepsilon}{m}(\beta - \cos 2t)(y_1 \cos mt + y_2 \sin mt) \sin mt \\[2mm] \dot{y}_2 &= \tfrac{\varepsilon}{m}(\beta - \cos 2t)\ (y_1 \cos mt + y_2 \sin mt) \cos mt. \end{aligned}$$

The case $m = 1$.

In example 11.4 we studied this case with $\beta = 0$ and we found instability. We apply the periodicity condition of Poincaré-Lindstedt to system $A2-1$ as in section 10.1. We find

$$\int_0^{2\pi} (\beta - \cos 2t)(y_1 \cos t + y_2 \sin t) \sin t \, dt = 0$$
$$\int_0^{2\pi} (\beta - \cos 2t)(y_1 \cos t + y_2 \sin t) \cos t \, dt = 0$$

and so to first order
$$y_2(0)(\beta + \tfrac{1}{2}) = 0$$
$$y_1(0)(\beta - \tfrac{1}{2}) = 0$$

so that $y_1(0) = 0$, $\beta = -\tfrac{1}{2}$ or $y_2(0) = 0, \beta = \tfrac{1}{2}$. Periodic solutions exist if $a = 1 \pm \tfrac{1}{2}\varepsilon + O(\varepsilon^2)$.

The case $m = 2$.

Applying the periodicity condition to system $A2 - 1$ produces

$$\int_0^\pi (\beta - \cos 2t)(y_1 \cos 2t + y_2 \sin 2t) \sin 2t = 0$$
$$\int_0^\pi (\beta - \cos 2t)(y_1 \cos 2t + y_2 \sin 2t) \cos 2t = 0$$

The calculation to first order yields $\beta = 0$; so we have to assume $\beta = O(\varepsilon)$ in system $A2 - 1$ after which we apply the periodicity condition to second order. After some calculations we find $\beta = \varepsilon/48$ or $\beta = -5\varepsilon/48$. Periodic solutions exist if

$$a = 4 - \frac{1}{48}\varepsilon^2 + O(\varepsilon^4)$$

or

$$a = 4 + \frac{5}{48}\varepsilon^2 + O(\varepsilon^4).$$

In figure $A2 - 1$ we have indicated the domains of stability of the trivial solution of the Mathieu-equation in the $a, \varepsilon$-parameter plane.

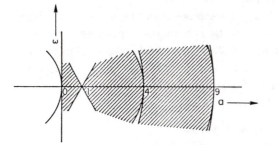

Figure A2-1. Domains of stability have been shaded

# Appendix 3: Trigonometric formulas and averages

$$\sin\alpha\cos\beta = \tfrac{1}{2}\sin(\alpha+\beta) + \tfrac{1}{2}\sin(\alpha-\beta)$$

$$\cos\alpha\cos\beta = \tfrac{1}{2}\cos(\alpha+\beta) + \tfrac{1}{2}\cos(\alpha-\beta)$$

$$\sin\alpha\sin\beta = \tfrac{1}{2}\cos(\alpha-\beta) - \tfrac{1}{2}\cos(\alpha+\beta)$$

$$\sin\alpha\cos\alpha = \tfrac{1}{2}\sin 2\alpha$$

$$\sin^2\alpha = \tfrac{1}{2} - \tfrac{1}{2}\cos 2\alpha$$

$$\cos^2\alpha = \tfrac{1}{2} + \tfrac{1}{2}\cos 2\alpha$$

$$\sin^3\alpha = \tfrac{3}{4}\sin\alpha - \tfrac{1}{4}\sin 3\alpha$$

$$\cos^3\alpha = \tfrac{3}{4}\cos\alpha + \tfrac{1}{4}\cos 3\alpha$$

$$\sin^4\alpha = \tfrac{3}{8} - \tfrac{1}{2}\cos 2\alpha + \tfrac{1}{8}\cos 4\alpha$$

$$\cos^4\alpha = \tfrac{3}{8} + \tfrac{1}{2}\cos 2\alpha + \tfrac{1}{8}\cos 4\alpha$$

$$\tfrac{1}{2\pi}\int_0^{2\pi}\sin^2\alpha\,d\alpha = \tfrac{1}{2\pi}\int_0^{2\pi}\cos^2\alpha\,d\alpha = \tfrac{1}{2}$$

$$\tfrac{1}{2\pi}\int_0^{2\pi}\sin^4\alpha\,d\alpha = \tfrac{1}{2\pi}\int_0^{2\pi}\cos^4\alpha\,d\alpha = \tfrac{3}{8}$$

$$\tfrac{1}{2\pi}\int_0^{2\pi}\sin^2\alpha\cos^2\alpha\,d\alpha = \tfrac{1}{8}$$

$$\tfrac{1}{2\pi}\int_0^{2\pi}\sin^6\alpha\,d\alpha = \tfrac{1}{2\pi}\int_0^{2\pi}\cos^6\alpha\,d\alpha = \tfrac{5}{16}$$

$$\tfrac{1}{2\pi}\int_0^{2\pi}\sin^4\alpha\cos^2\alpha\,d\alpha = \tfrac{1}{2\pi}\int_0^{2\pi}\sin^2\alpha\cos^4\alpha\,d\alpha = \tfrac{1}{16}$$

# Answers and hints to the exercises

**2.1** Putting $x_1 = x$, $x_2 = \dot{x}$ we have the system

$$\dot{x}_1 = x_2$$
$$\dot{x}_2 = \lambda x_2 + (\lambda - 1)(\lambda - 2)x_1.$$

If $\lambda \neq 1$ and $\lambda \neq 2$, $(0,0)$ is critical point. If $\lambda = 1$ or $\lambda = 2$, the $x_1$-axis consists of critical points. The matrix $\partial f/\partial x$ from section 2.2 is in all cases

$$\begin{pmatrix} 0 & 1 \\ (\lambda - 1)(\lambda - 2) & \lambda \end{pmatrix}$$

If $\lambda = 1$ or $\lambda = 2$ the matrix is singular, the critical points are degenerate. If $\lambda \neq 1$ and $\lambda \neq 2$ the eigenvalues are

$$\frac{1}{2}\lambda \pm \frac{1}{2}\sqrt{5(\lambda^2 - 12 + 8)}.$$

$\lambda < 1$ and $\lambda > 2$: $(0,0)$ is a saddle;
$1 < \lambda < 2$ : $(0,0)$ is a node with negative attraction.

**2.2** Critical points are $(0,0)$ and $(x_0, y_0) = \left(\frac{c}{b-cs}, \frac{a}{b-cs}\right)$ with $b - cs > 0$. As in example 2.7 $(0,0)$ is a saddle point. Linearising near the second critical point we find

$$\dot{x} = s\frac{ca}{b}(x - x_0) - c(y - y_0) + \dots$$
$$\dot{y} = (a - s\frac{ca}{b})(x - x_0) + \dots$$

Analysis of the linearised system for $b - cs > 0$ yields that $(x_0, y_0)$ is a negative attractor.

**2.3**  a. 5 critical points: $(0,0,0), (\pm 1, 0, 1), (\frac{1}{2}, \pm\frac{1}{2}\sqrt{3}, 1)$.

  b. Putting $x_3(t) = 0$ or $x_3(t) = 1$ solves the equation for $x_3$.

  c. The equations describing the flow in the set $x_3 = 1$ are

$$\dot{x}_1 = x_1^2 + x_2^2 - 1$$
$$\dot{x}_2 = x_2(1 - 2x_1).$$

  Analysis of the directions of the phase-flow shows that no periodic solutions exist in this set.

**2.4**  b. Putting $p = m\dot{x}, q = x$ we have

$$H(p,q) = \frac{1}{2}\frac{p^2}{m} + \frac{1}{2}kq^2 + kln(a-q), q < a.$$

c. In vector form we have with $x = x_1, \dot{x} = x_2$ critical points if

$$x_2 = 0 \ , \ x_1^2 - ax_1 + \lambda = 0$$

or

$$x_1 = \frac{1}{2}a \pm \frac{1}{2}\sqrt{a^2 - 4\lambda}.$$

$0 < \lambda < \frac{a^2}{4}$ :  two critical points, the minus sign yields a centre, the plus sign a saddle point.

$\lambda = \frac{a^2}{4}$:  the two critical points coalesce to a degenerate critical point.

$\lambda > \frac{a^2}{4}$ :  there are no critical points and so no equilibrium positions.

In the case $0 < \lambda < a^2/4$, an infinite set of periodic solutions exists around the centre point, corresponding with oscillations of the small conductor. Starting for $\dot{x} = 0$ and $x(t_0)$ between the saddle point and $x = a$, the small conductor moves to the long conductor.
The last behaviour holds for all initial conditions if $\lambda > a^2/4$.

If $\lambda < 0$, which is achieved when the two currents move in opposite directions, there is one critical point (for $x < a$):

$$x_2 = 0, x_1 = \frac{1}{2}a - \frac{1}{2}\sqrt{a^2 - 4\lambda}.$$

This is a centre; outside the critical point all motions are periodic. Mechanically this means that the conductors are repulsive, the spring produces a counter-acting force.

The non-degenerate critical points are either centres or saddles; this is in agreement with theorem 2.1.

**2.5**  $(0,0)$ saddle, $(\pm\frac{1}{2}\sqrt{2}, 0)$ both centres by linear analysis. The saddle cannot be an attractor for the full nonlinear system. Note that we can write

$$\frac{dx}{dy} = \frac{y}{x - 2x^3}$$

and by separation of variables we find the first integral $x^2 - x^4 - y^2 =$ constant. Applying the Morse lemma (2.3) we find that $(\pm\frac{1}{2}\sqrt{2}, 0)$ are centres in the full nonlinear system (no attraction).

2.6   a. $F = y - cx$ with $c$ a constant.

   b. No, the flow is not volume(area)-preserving: the divergence is 2, $(0,0)$ is a negative attractor.

2.7 Consider in $\mathbb{R}^2$ $\dot{x} = f(x,y), \dot{y} = g(x,y)$. The phase-flow is described by the equation

$$f(x,y)\frac{dy}{dx} - g(x,y) = 0.$$

The equation is exact, i.e. the solutions are described by the implicit relation $F(x,y) = c$ if

$$\frac{\partial}{\partial x}f(x,y) = \frac{\partial}{\partial y}(-g(x,y)).$$

This condition implies that the flow is divergence free.

2.8 The critical points are $(0,0), (1,1), (1,-1)$. The divergence of the vector field is zero, so there are no attractors. The equation for $dy/dx$ is exact (exercise 2.7) with first integral $F(x,y) = x^2y - \frac{1}{2}x^4 - \frac{1}{4}y^4$. Analysis of the phase-orbits described by $F(x,y) = $ constant shows that an infinite number of periodic solutions exist.

3.1 There are five critical points:
   $(0,0)$      saddle point;
   $(1,0)$      focus, negative attractor;
   $(-1.0)$   degenerate, eigenvalues 0 and 1;
   $(0,1)$      focus, negative attractor;
   $(0,-1)$   degenerate, eigenvalues 0 and 1.

3.2   a. Introduce polar coordinates $x = r\cos\theta, y = r\sin\theta$.
       We find

$$\begin{aligned} r\dot{\theta} &= r(c\cos^2\theta + d\sin\theta\cos\theta - a\sin\theta\cos\theta - b\sin^2\theta) + o(r) \text{ as } r \to 0 \\ &= r\beta(t) + o(r). \end{aligned}$$

As we have a focus by linearisation $\int_0^t \beta(\tau)d\tau \to +\infty$ or $-\infty$ as $t \to \infty$. This remains the dominating term as $r \to 0$.

   b. No.
       In this special case we have a node by linearisation. In polar coordinates we have for the full system

$$\dot{\theta} = 1/\log r , \quad \dot{r} = -r$$

so that $r(t) = r(0)e^{-t}$

$$\theta(t) = \theta(0) + \log(-\log r(0)) - \log(t - \log r(0)).$$

Spiralling behaviour.

3.3  a. $(-1,-1)$ saddle, no attraction; $(+1,+1)$ saddle, no attraction; $(-1,+1)$ focus, positive attractor; $(+1,-1)$ focus, negative attractor.

3.4  a. One critical point : $(0,0)$, a centre.

 b. The equation for $r = \sqrt{(x^2 + y^2)}$ is $\dot{r} = r^3 \sin r$. So the solution $(0,0)$ is a negative attractor.

3.5 In vector form the equation becomes with $(x, \dot{x}) = (x_1, x_2)$

$\dot{x}_1 = x_2$

$\dot{x}_2 = x_1(1 - x_1) - cx_2.$

Critical points are $(0,0)$ with eigenvalues $-c \pm \sqrt{c^2 + 4}$ and $(1,0)$ with eigenvalues $-c \pm \sqrt{c^2 - 4}$.

For $(1,0)$ to be a positive attractor, $(0,0)$ a negative attractor we have to take $c > 0$. If $0 < c < 2$, the orbits near $(1,0)$ are spiralling; to have monotonic behaviour of $x(t)$ a necessary condition is $c \geq 2$. Note that if solutions with the required properties exist, they will correspond with one of the unstable manifolds of $(0,0)$ which coincides with one of the stable manifolds of $(1,0)$.

3.6 The solutions $(\varphi(t), 0)$ with $\dot{\varphi} = \varphi^3$ tend to $(0,0)$ as $t \to -\infty$. However, we find solutions with the property $\dot{x} < 0, \dot{y} > 0$ when crossing the negative $y$-axis arbitrarily close to $(0,0)$. So $(0,0)$ is not an attractor.

3.7 There are nine critical points:

$(0,0)$ node, negative attractor;

$(0, \pm 1/\sqrt{3})$ 2 nodes, positive attractors;

$(\pm 1, 0)$ 2 nodes, positive attractors;

$(\pm 1/\sqrt{5}, \pm \sqrt{2}/\sqrt{15})$ 4 saddle points.

4.1 The divergence of the vector function describing the phase-flow is 1. It follows from Bendixson's criterion that no periodic solutions exist.

4.2 Applying theorem 4.6 we note that

$$F(x) = ax + \frac{1}{2}bx^2 + \frac{1}{3}cx^3 + \frac{1}{4}dx^4.$$

So $b = d = 0$, $c > 0$, $a < 0$. As $\alpha = \beta$ (notation of theorem 4.6), there is only one limit cycle in this case.

4.3    a. Applying Bendixson's criterion we find for the divergence $ky^{k-1}$ : no periodic solutions if $k$ is odd.

      b. $(0,0)$ is a centre by linear analysis. The expression $(y^2 - \frac{1}{2} + x)e^{2x}$ is a first integral; applying the Morse lemma (2.3) we find periodic solutions (cycles), no limit cycles.

4.4 Differentiation of the Rayleigh equation produces

$$\ddot{x} + \dot{x} = \mu(1 - 3\dot{x}^2)\ddot{x}$$

We obtain the van der Pol equation by putting $y = \dot{x}\sqrt{3}$.

4.5 The divergence of the vector function is $y + x$, which is sign definite apart from $(0,0)$. So there are no periodic solutions because of Bendixson's criterion (thereom 4.1).

4.6    a. The divergence of the vector field is $-3x^2 + \mu$ so there are no periodic solutions if $\mu < 0$ (Bendixson). With $r^2 = x^2 + y^2$ we derive

$$\dot{r} = -\frac{1}{r}x^2(x^2 - \mu),$$

so if $\mu = 0, \dot{r} \le 0$; the phase flow is contracting except when passing the $y$-axis. There is no periodic solution if $\mu = 0$. To discuss $\mu > 0$ we differentiate the equation for $x$ to obtain

$$\ddot{x} + x = (\mu - 3x^2)\dot{x}$$

to obtain the Liénard equation 4.5 or we apply directly theorem 4.6 as the system is of the form 4.6. $\alpha = \beta = \sqrt{\mu}$ so there exists one periodic solution.

      b. As $\mu \downarrow 0$, the construction in the proof of theorem 4.6 shows that the limit cycle contracts around $(0,0)$ and vanishes into $(0,0)$ as $\mu = 0$.

4.7 The orbit corresponding with $\varphi(t)$ is $\gamma$ and we have $\omega(\gamma) = a$. According to theorem 4.2 $\omega(\gamma)$ is closed and invariant, so $x(t) = a$ is a solution.

4.8    a. If $\lambda = 0$ the system has a Hamilton function: $E(x,y)$. Critical points $(0,0)$ saddle, $(\pm1,0)$ centres. The orbits are described by $E(x,y) = $ constant; the saddle remains a saddle, the centres remain centres (see figure).

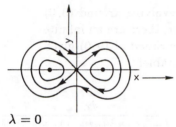

$\lambda = 0$

b. If $\lambda \neq 0$ the linear analysis doesnot change. We compute:

$$L_t E = \lambda E [(\frac{\partial E}{\partial x})^2 + (\frac{\partial E}{\partial y})^2].$$

So $E(x, y) = y^2 - 2x^2 + x^4 = 0$ represents an invariant set. The saddle loops $\beta_1$ and $\beta_2$ persist.

Choose $\lambda > 0$. Starting inside a saddle loop we have $E < 0$, $L_t E < 0$, so the orbits spiral towards the critical points $(1, 0)$ or $(-1, 0)$. Outside the saddle loops we have $E > 0$, $L_t E > 0$ so the orbits move away from $\beta_1$ and $\beta_2$; see the figure

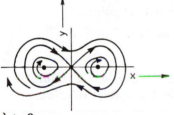

$\lambda > 0$

If $\lambda < 0$ we find in the same way that $(1, 0)$ and $(-1, 0)$ are negative attractors.

c. $\omega(\gamma_1^+) = \beta_1 \cup (0, 0); \omega(\gamma_2^+) = \beta_2 \cup (0, 0);$
$\omega(\gamma_3^+) = \beta_1 \cup (0, 0) \cup \beta_2.$

5.1   a. Stable.

    b. $\alpha < 0$ unstable, $\alpha = 0$ Lyapunov-stable, $\alpha > 0$ asymptotically stable.

    c. stable (see example 4.6).

5.2 Introducing polar coordinates $x = r \cos \theta, y = r \sin \theta$ we find

$$\dot{r} = r \sin^2 \theta \sin^2 \left(\frac{\pi}{r^2}\right), r > 0$$
$$\dot{\theta} = \frac{1}{2} \sin(2\theta) \sin^2 \left(\frac{\pi}{r^2}\right) - 1.$$

So we have $\dot{r} \geq 0, \dot{\theta} \leq -1/2$; the orbits keep revolving around $(0,0)$ and they are "repelled" from the origin. However, there are an infinite number of unstable limit cycles in each neighbourhood of $(0,0)$ given by $r = 1/\sqrt{n}, n = 1, 2, 3, \ldots$ $(0,0)$ is Lyapunov-stable!

5.3 We expand

$$
\begin{aligned}
T &= 4 \int_0^a \frac{dx}{(2 \cos x - 2 \cos a)^{1/2}} = 4 \int_0^a \frac{dx}{(a^2 - x^2 - \frac{1}{12}(a^4 - x^4) + \cdots)^{1/2}} \\
&= 4 \int_0^a \frac{dx}{(a^2 - x^2)^{1/2}}(1 + \frac{1}{24}(a^2 + x^2) + \ldots) \\
&= 2\pi(1 + \frac{1}{16}a^2 + O(a^4))
\end{aligned}
$$

For the last step we used an integral table. This expansion also shows why the harmonic oscillator is a good approximation of the mathematical pendulum if the amplitude of oscillation is not too big.

5.4  a. Critical points exist if $b^2 - 4c \geq 0$. Linear analysis produces the eigenvalues $\pm(b^2 - 4c)^{1/4}, \pm i(b^2 - 4c)^{1/4}$. If $b^2 - 4c > 0$: a saddle and a centre. If $b^2 - 4c = 0$ : degenerate. The saddle point is unstable, the centre is also a centre for the full nonlinear system (cf. example 2.14) and is Lyapunov-stable. If $b^2 - 4c = 0$ there is one unstable critical point.

b. Around the centre, $b^2 - 4c > 0$, there exists a family of periodic solutions. These are stable.

5.5 Linearisation produces the eigenvalues $-1, 1 - a$. We have $a > 1$ stability, $a < 1$ instability, $a = 1$ instability (consider the special solutions corresponding with $x(t) = y(t)$).

6.1 The eigenvalues of matrix $A$ are $-3$ (multiplicity 1) and $-1$ (multiplicity 2). Application of theorem 6.3 yields asymptotic stability of $x = 0$.

6.2 We split $A(t) = B + C(t)$:

$$
A(t) = \begin{pmatrix} 0 & 1 & 0 \\ 0 & 0 & 1 \\ a-2 & -1 & -a \end{pmatrix} + \begin{pmatrix} e^t & \frac{1}{t^2} & e^{-t} \\ \frac{\sin t}{t^{3/2}} & 0 & e^{-t} \\ 2 - a & 0 & \frac{a}{t} \end{pmatrix}
$$

The equation for the eigenvalues of $B$ is $\lambda^3 + a\lambda^2 + \lambda + 2 - a = 0$ with solutions $-1$ and

$$
\frac{1}{2}(1 - a) \pm \frac{1}{2}(a^2 + 2a - 7)^{1/2}.
$$

We note that $C(t)$ satisfies the condition $\lim_{t \to \infty} \|C(t)\| = 0$.

If the real parts of the eigenvalues of matrix $B$ are negative, theorem 6.3 applies; so we have asymptotic stability of $x = 0$ if $1 < a < 2$. Theorem 6.4 yields that $x = 0$ is unstable if $a < 1$ or $a > 2$.

6.3    a. Using variation of constants we have (cf. the proof of theorem 6.2)

$$x(t) = \phi(t)x_0 + \int_0^t \phi(t-\tau)B(\tau)x(\tau)d\tau + \int_0^t \phi(t-\tau)C(\tau)d\tau$$

where $\phi(t)$ is the fundamental matrix solving $\dot{y} = Ay$, $\phi(0) = I$. We have

$$\|\phi(t)\| \le c_3 , \ c_3 \text{ a positive constant}$$

so that

$$\|x(t)\| \le c_3\|x_0\| + \int_0^t c_3\|B(\tau)\|\|x(\tau)\|d\tau + c_3 c_2.$$

Applying Gronwall's inequality yields

$$\|x(t)\| \le (c_3\|x_0\| + c_3 c_2)e^{c_3 \int_0^t \|B(\tau)\|d\tau} \le (c_3\|x_0\| + c_3 c_2)e^{c_3 c_1}.$$

   b. No.

   c. We can for instance assume $\|C(t)\|$ is uniformly bounded, but then we should add

$$Re(\lambda_i) < 0, \ i = 1,\ldots,n$$

to obtain the boundedness of $x(t)$.

6.4 Using variation of constants as in the proof of theorem 6.3 we find from the estimate 6.7

$$\|x\| \le C_1 e^{-\mu t}\|x_0\| + \int_0^t C_1 b e^{-\mu(t-\tau)}\|x\|d\tau$$

with $C_1$ and $\mu$ positive constants. Multiplying with exp.$(\mu t)$ and applying Gronwall's lemma (1.2) we find

$$\begin{aligned} \|x\|e^{\mu t} &\le C_1\|x_0\|e^{C_1 bt} \text{ or} \\ \|x\| &\le C_1\|x_0\|e^{(C_1 b-\mu)t} \ ,t \ge 0. \end{aligned}$$

Choosing $0 < b < \mu/C_1$ yields the required result.

6.5 $(0,0)$ is unstable as $c(e^{2t}\sin t, e^{2t}\cos t)$ is a solution with $c$ arbitrary small.

From theorem 6.6 we have for the characteristic exponents $\lambda_1 + \lambda_2 = \frac{1}{2\pi}\int_0^{2\pi}(2 + \cos t + 2 + \sin t)dt = 4$. As $\lambda_1 = 2$, $\lambda_2 = 2$ no nontrivial solution with the required property exists.

6.6 According to theorem 6.5 we can write the fundamental matrix as $P(t)$ $\exp(Bt)$ with $P(t)$ $2\pi$-periodic and $B$ a constant $2 \times 2$-matrix. For the characteristic exponents $\lambda_1, \lambda_2$ we have with theorem 6.6

$$\lambda_1 + \lambda_2 = \frac{1}{2\pi} \int_0^{2\pi} SpA(t)dt = 2$$

so at least one characteristic exponent is positive.

6.7 a. The solution is $x(t) = x_0 e^{\int_0^t f(\tau)d\tau}$. Introduce $f^0 = \frac{1}{T}\int_0^T f(t)dt$; the solution can be written as

$$x(t) = x_0 e^{\int_0^t (f(\tau)-f^0)d\tau+f^0 t}.$$

From the Fourier expansion of $f(t)$ (or in a more elementary way) we know that $\int_0^t (f(\tau) - f^0)d\tau$ is $T$-periodic, so $B = f^0$ and $P(t) = \exp(\int_0^t (f(\tau) - f^0)d\tau)$.

b. In both cases the condition is $f^0 = 0$.

c. The solution is

$$\begin{pmatrix} x_1(t) \\ x_2(t) \end{pmatrix} = e^{\int_0^t f(\tau)d\tau \begin{pmatrix} a & b \\ c & d \end{pmatrix}} \begin{pmatrix} x_1(0) \\ x_2(0) \end{pmatrix}$$

$$P(t) = \exp\left(\int_0^t (f(\tau) - f^0)d\tau \begin{pmatrix} a & b \\ c & d \end{pmatrix}\right), B = e^{f^0 \begin{pmatrix} a & b \\ c & d \end{pmatrix}}.$$

d. For the solution to be bounded as $t \to \pm\infty$ the real parts of the eigenvalues of $B$ have to be zero. So $f^0 = 0$ or $a+d = 0$, $ad - bc \geq 0$.
For the solutions to be periodic they have to be bounded as $t \to \pm\infty$ and the ratio of the imaginary parts of the eigenvalues and $T$ must be rational.

e. No. To see this write $x = (x_1, x_2)$ and $z = x_1 + ix_2$. The complex variable $z$ satisfies the equation $\dot{z} = \exp(it)\bar{z}$ with the solution $z(t) = (p + qi)\exp((\mu + i\nu)t)$; determine the constants by substitution.

7.2 a. $(0,0)$ is an unstable focus; putting the coefficients of $\dot{x}$ equal to zero produces an orbit of the harmonic oscillator equation, so $x(t) = \cos t$, $\dot{x}(t) = \sin t$ is a periodic solution. Transforming to polar coordinates also produces this result: put $x = r\cos\theta$, $\dot{x} = r\sin\theta$ so that

$$\dot{r} = (1 - r^2)r\sin^2\theta \ , \ \dot{\theta} = -1 + (1 - r^2)\sin\theta\cos\theta.$$

b. Stability and limit cycle character of the periodic solution can also be obtained in two ways. First, it is clear that if $r$ is near 1, $\theta(t)$ is monotonic and $\lim_{t \to \infty} r(t) = 1$.

Secondly, as we know the periodic solution explicitly, it is easy to put $x = \cos t + y$, $\dot{x} = -\sin t + \dot{y}$ to find

$$\ddot{y} + 2\sin^2 t\, \dot{y} - 2\sin t \cos t\, y = \dots$$

where the dots represent nonlinear terms.

Applying theorem 7.4 we find (cf. example 7.4) one characteristic exponent zero and for the other one (theorem 6.6) $-1$. So we have stability.

7.3   a. $(\frac{1}{2}, -\frac{1}{2})$ centre, $(-1, 1)$ saddle.

b. The divergence of the vector function on the righthand side is $-2x - 2y$. Use the reasoning of Bendixson (theorem 4.1).

c. Putting $x = r \sin \varphi, y = r \cos \varphi$ we find

$$\dot{r} = (\sin \varphi + \cos \varphi)(1 - r^2)$$

$r = 1$ corresponds with a periodic solution: $x = \sin t$, $y = \cos t$.

d. Putting $x = \sin t + x_1, y = \cos t + y_1$ we find

$$\dot{x}_1 = -2 \sin t x_1 - 2 \cos t y_1 + y_1 + \dots \text{ (nonlinear terms)}$$
$$\dot{y}_1 = -x_1 - 2 \sin t x_1 - 2 \cos t y_1 + \dots$$

With example 7.4 and theorem 6.6 we find both characteristic exponents to be zero. So we cannot apply theorem 7.4 to establish stability.

e. Putting $x + y = u, x - y = v$ we find

$$\dot{u} = 2 - v - u^2 - v^2$$
$$\dot{v} = u$$

and with $u^2 = w$

$$\frac{1}{2} \frac{dw}{dv} = 2 - v - v^2 - w$$

The equation for $w$ is linear in $w$ and can be integrated to produce a first integral of the system. Applying the Morse lemma 2.3 we find that $x = \frac{1}{2}$, $y = -\frac{1}{2}$ is a centre for the system. The periodic solution found in c is stable, it is one of an infinite set of cycles around the centre.

7.4    a. $(0,0)$ node, negative attractor;

If $a \neq 1$ there are three other isolated critical points:

$(0,1)$ if $0 < a < 1$ saddle, if $a > 1$ attracting node;

$(1,0)$ if $0 < a < 1$ saddle, if $a > 1$ attracting node;

$(\frac{1}{1+a}, \frac{1}{1+a})$ if $0 < a < 1$ attracting node, if $a > 1$ saddle.

If $a = 1$ all points satisfying $x + y = 1, x \geq 0, y \geq 0$ are critical; one eigenvalue is zero, the other is $-1$.

   b. Saddles are not attracting, the attraction properties of the nodes carry over to the nonlinear system.

If $a = 1$, outside $x = 0, y = 0$ and outside the critical line $x+y = 1$ the flow is described by the equation $dy/dx = y/x$. So the phase orbits are straight lines attracting towards the critical line $x+y = 1$; note that the critical points on $x+y = 1$ are not attracting, however the critical line is an attracting set.

   c. The boundary $x = 0, y = 0$ consists of phase orbits. Apply the uniqueness theorem.

d-e Note that the line $y = x$ consists of phase orbits for all $a > 0$.

$\lim\limits_{t \to \infty} x(t) = 0$ if $a > 1$ and $y(0) > x(0)$;

$\lim\limits_{t \to \infty} y(t) = 0$ if $a > 1$ and $y(0) < x(0)$;

extinction of both species doesnot occur;

   coexistence   if $0 < a \leq 1$ and all positive initial values,
   if $a > 1$ if and only if $x(0) = y(0)$

7.5 There are 7 critical points; the stability of 5 of them can be determined by linear analysis:

$(1,0,0)$ eigenvalues $0, 4, 8$, unstable;

$(0,1,0)$ eigenvalues $0, 4\sqrt{2}, -4\sqrt{2}$, unstable;

$(0,-1,0)$ eigenvalues $0, 4\sqrt{2}, -4\sqrt{2}$, unstable;

$(1,1,-2)$ eigenvalues $32, 12\sqrt{3}, -12\sqrt{3}$, unstable;

$(\frac{3}{5}, \frac{1}{5}, 2)$ eigenvalues $-\frac{32}{5}, -8 + 4i, -8 - 4i$, asymptotically stable.

The two degenerate points are discussed as follows.

$(1,0,0)$ eigenvalues $0, -4, -8$; note that $z = 0$ is an invariant plane, the solutions with $-2 < z < 0$ approach this plane, the solutions with $z > 0$ are leaving a neighbourhood of the plane so we have instability.

$(0,0,1)$ eigenvalues $0, 0, 0$; for $-2 < z < +2$ we have $\dot{z} > 0$ if $x \neq 0, y \neq 0$: instability.

7.6    a. If $n = 1$ the eigenvalues are $\alpha \pm i$, if $n \geq 2$ they are $\alpha, \alpha$; so the instability if $\alpha > 0$, asymptotic stability if $\alpha < 0$.

b. If $\alpha = 0$ $dx/dy = -y^n/x^n$ which admits the first integral $x^{n+1} + y^{n+1} =$ constant. If $n$ is odd, the phase orbits are closed curves around $(0,0)$: Lyapunov stability. If $n$ is even one can analyse the direction field of the flow to find instability; one can also note that $x = -y$ consists of three phase orbits one of which approaches $(0,0)$, another leaves a neighbourhood of $(0,0)$.

7.7   a. $(0,0,0)$ , three real eigenvalues, 1 positive, 2 negative, unstable;
$(c,0,a)$, two eigenvalues purely imaginary, one real $(c-b)$
$(b,a,0)$, two eigenvalues purely imaginary, one real $(b-c)$.

b. Families of periodic solutions exist in the $x,y$- and $x,z$-coordinate planes. Take now $y(0)z(0) > 0$. The last two equations produce

$$ z\dot{y} - y\dot{z} = (c-b)yz $$

or $\frac{d}{dt}\left(\frac{y}{z}\right) = (c-b)\frac{y}{z}$
which can be integrated and yields

$$ y(t) = z(t)\frac{y(0)}{z(0)}e^{(c-b)t}. $$

If $c \neq b$ there are no periodic solutions with $y(t)z(t) > 0$.
If $c = b$ and putting $y(0)/z(0) = \alpha$ we have $y(t) = \alpha z(t)$

$$ \dot{x} = ax - (\alpha+1)xz $$
$$ \dot{z} = -cz + xz $$

In this case all solutions with $x(0)y(0)z(0) > 0$ are periodic.

c. This is possible if $b \neq c$. For instance if $b < c$, $(b,a,0)$ has a one-dimensional stable manifold. Starting on this manifold we have $\lim_{t\to\infty} z(t) = 0$.

8.1  Inspired by the case $f = 0$ we try as a Lyapunov function

$$ V(x,y) = \frac{1}{2}y^2 - \cos x + 1 $$

$V(x,y)$ is positive definite in a neighbourhood of $(0,0)$.
$L_t V(x,y) = y\dot{y} + \sin x\dot{x} = \sin x f(x,y)$
We choose $f(0,0) = 0$ and $\begin{array}{ll} f(x,y) \leq 0 & \text{if} \quad x > 0 \\ f(x,y) \geq 0 & \text{if} \quad x < 0, \end{array}$ for instance $f(x,y) = -x$ or $-x(1+y^2)$ or $-x\sin^2 y$ etc. Apply theorem 8.1.

8.2 No.  Consider example 6.2 and carry out Liouville-transformation to find a counter-example.

8.3 $(0,0)$ is Lyapunov-stable if $n$ is odd (theorem 8.5); $(0,0)$ is unstable if $n$ is even (for instance from the tangent vector field near $(0,0)$).

8.4   a. The first two equations produce the integral $x^2 + 2y^2$ (divide $\dot{x}/\dot{y} = dx/dy = \ldots$ and separate the variables), the last two equations produce the integral $y^2 + (z-1)^2$. Combination of the two integrals yields the Lyapunov function

$$V = x^2 + 3y^2 + z^2 - 2z.$$

$V$ is positive definite and $L_t V = 0$.

  b. No, $(0,0,z_0)$ is a solution for all $z_0$.

8.5 Choose the Lyapunov function $V = x^2 - 2y^2$; $L_t V = 2x^4 + 4y^6$. Application of theorem 8.3 shows that $(0,0)$ is unstable.

8.6   a. No.

  b. Choose as a possible Lyapunov function

$$V(t,x) = \dot{x}^2 + \phi(t)x^2$$

There exists a $t = t_0$ such that $0 < c/2 \le \phi(t)$ for $t \ge t_0$. So $V(t,x)$ is positive definite for $t \ge t_0$.

$$
\begin{aligned}
L_t V &= 2\dot{x}\ddot{x} + \dot{\phi}(t)x^2 + \phi(t)2x\dot{x} \\
&= \dot{\phi}(t)x^2.
\end{aligned}
$$

If $\phi(t)$ is monotonically decreasing we have $\dot{\phi}(t) \le 0$ so that $L_t V \le 0$; in this case theorem 8.1 applies and $(0,0)$ is stable. If $\phi(t)$ is monotonically increasing we have $\dot{\phi}(t) \ge 0$. We choose for $t \ge t_0$

$$U(t,x) = \frac{1}{\phi(t)}\dot{x}^2 + x^2$$

$U(t,x)$ is positive definite for $t \ge t_0$ and

$$L_t U = -\frac{\dot{\phi}}{\phi(t)}\dot{x}^2$$

so $L_t U \le 0$ if $\phi(t)$ is monotonically increasing, $(0,0)$ is stable according to theorem 8.1.

**8.7** It is no restriction to assume that $A$ is in diagonal form (if it is not, the assumptions on $A$ guarantee the existence of a suitable transformation). Choose

$$V = x_1^2 + \ldots + x_p^2 - x_{p+1}^2 - \ldots - x_n^2.$$

We have $L_t V = 2(\lambda_1 x_1^2 + \ldots + \lambda_p x_p^2 - \lambda_{p+1} x_{p+1}^2 - \ldots - \lambda_n x_n^2) + 2x.f(x)$. $L_t V$ is positive definite in a neighbourhood of $x = 0$, $V$ assumes positive values in each neighbourhood of $x = 0$. Theorem 8.3 yields instability.

**8.8** We shall try a quadratic Lyapunov-function. First put $\dot{x} = y$, $\dot{y} = z$, $\dot{z} = -f(y)z - ay - bx$ and take

$$V(x, y, z) = \frac{1}{2a}(az + by)^2 + \frac{1}{2}(bx + ay)^2 + b \int_0^y (f(s) - \frac{b}{a})s\, ds$$

and $L_t V = -a(f(y) - \frac{b}{a})z^2$
If $f(\dot{x}) \geq c > b/a$, $V$ is positive definite and $L_t V \leq 0$. Outside the plane $z = 0$ we have $L_t V < 0$, moreover the plane $z = 0$ $(x, y, z) \neq (0, 0, 0)$, is transversal to the flow. The same reasoning which led us to the corollary of theorem 8.2 yields that $(0, 0, 0)$ is a global attractor.

**8.9** We try a quadratic Lyapunov-function $V = ax^2 + by^2$. Choosing $a = 1, b = 2$ we find $L_t V < 0$; $(0, 0)$ is asymptotically stable.

**9.1** We put $x(t) = x^{(0)}(t) + \varepsilon x^{(1)}(t) + \varepsilon^2 \ldots$
Substitution and equating equals powers of $\varepsilon$ yields

$$\begin{aligned}
x^{(0)}(t) &= \cos t \\
x^{(1)}(t) &= \tfrac{3}{8} t \sin t + \tfrac{1}{4} \sin t \sin 2t + \tfrac{1}{32} \sin t \sin 4t + \\
&\quad + \tfrac{1}{4} \cos t \cos^4 t - \tfrac{1}{4} \cos t.
\end{aligned}$$

Note that, as $x^{(0)}(t)$ satisfies the initial conditions, we choose $x^{(1)}(0) = \dot{x}^{(1)}(0) = 0$,

**9.2**   a. The solution is expanded as

$$x(t) = x^{(0)}(t) + \varepsilon x^{(1)}(t) + \ldots$$

$$y(t) = y^{(0)}(t) + \varepsilon y^{(1)}(t) + \ldots$$

Substitution in the equations and equating equal powers of $\varepsilon$ produces

$$x^{(0)}(t) = \frac{x_0 e^t}{1 - x_0 + x_0 e^t} \qquad y^{(0)}(t) = \frac{y_0 e^t}{1 - y_0 + y_0 e^t}$$

$$x^{(1)}(t) = \frac{e^t}{(1 - x_0 + x_0 e^t)^2} [\frac{x_0}{y_0}(x_0 - y_0) \log(y_0 e^t + 1 - y_0)$$
$$-x_0^2 e^t + x_0^2]$$

and a similar expression for $y^{(1)}(t)$ (interchange $x_0$ and $y_0$).

b. We find

$$\lim_{t \to \infty}(x^{(0)}(t) + \varepsilon x^{(1)}(t)) = \lim_{t \to \infty}(y^{(0)}(t) + \varepsilon y^{(1)}(t)) = 1 - \varepsilon$$

The expansion is valid for $0 \le t \le h$ (an $\varepsilon$-independent constant) and for $t$ in a neighbourhood of $\infty$. Here we cannot conclude anything about the uniform validity of the expansion. However, in chapter 4 of Sanders and Verhulst (1985) it is shown that the approximation is valid for all time.

9.3   a. According to theorem 9.1: $\|x(t) - y(t)\| = O(\varepsilon^2)$ on the time-scale 1. Rewriting the problems as integral equations for $x$ and $y$ we have

$$x(t) = \eta + \varepsilon \int_{t_0}^t f_1(\tau, x(\tau)) d\tau + \varepsilon^2 \int_{t_0}^t f_2(\tau, x(\tau)) d\tau$$
$$y(t) = \eta + \varepsilon \int_{t_0}^t f_1(\tau, y(\tau)) d\tau.$$

Subtraction yields

$$x(t) - y(t) = \varepsilon \int_{t_0}^t [f_1(\tau, x(\tau)) - f_1(\tau, y(\tau))] d\tau + \varepsilon^2 \int_{t_0}^t f_2(\tau, x(\tau)) d\tau$$

Using the Lipschitz-continuity of $f_1$ and the boundedness of $f_2$ we find

$$\|x(t) - y(t)\| \le \varepsilon L \int_{t_0}^t \|x(\tau) - y(\tau)\| d\tau + \varepsilon^2 M(t - t_0).$$

We apply Gronwall's inequality in the form 1.3 with $\delta_2 = \varepsilon^2 M, \delta_1 = \varepsilon L, \delta_3 = 0$:

$$\|x(t) - y(t)\| \le \varepsilon \frac{M}{L} e^{\varepsilon L(t - t_0)} - \varepsilon \frac{M}{L} \quad , t \ge t_0.$$

So $\|x(t) - y(t)\| = O(\varepsilon)$ on the time-scale $1/\varepsilon$.

b. According to theorem 9.1: $\|x(t) - y(t)\| = O(\varepsilon^3)$ on the time-scale 1. In an analogous way as before we find

$$\|x(t) - y(t)\| \le \varepsilon^2 L \int_{t_0}^t \|x(\tau) - y(\tau)\| d\tau + \varepsilon^3 M(t - t_0).$$

Again using theorem 1.3 we have

$$\|x(t) - y(t)\| \le \varepsilon \frac{M}{L} e^{\varepsilon^2 L(t-t_0)} - \varepsilon \frac{M}{L} \ , t \ge t_0.$$

So $\|x(t) - y(t)\| = O(\varepsilon)$ on the time-scale $1/\varepsilon^2$.
Note that on enlarging the time-scale, we are too pessimistic about the accuracy on a smaller time-scale, of course this is only natural.

9.4   a. $(x, \dot{x}) = (0,0)$ is a centre, $(1/\lambda, 0)$ a saddle point. The periodic solutions are located around the centre and within the saddle loop.

b. If $0 < \varepsilon \ll 1$ $(0,0)$ is an unstable focus, $1/\lambda, 0)$ a saddle point.

c. According to the expansion theorem 9.1 the solutions are $\varepsilon$-close to the solutions of the unperturbed problem on the time-scale 1. So periodic solutions have to be found in the region within the saddle loop (with $O(\varepsilon)$ error) of the unperturbed equation.

d. The Bendixson criterion implies that a periodic solution, if it exists, must intersect one or both of the lines $x = \pm 1$. For the saddle loop to require that solutions in the interior intersect $x = \pm 1$ we have $\lambda < 1$.

10.1   a. $(0,0)$ is a saddle, $(-1,0)$ a centre; compare with figure 2.10 which is related.

b. We translate $x = -1 + \varepsilon y$ to find

$$\ddot{y} + y = \varepsilon y^2 \ , \ y(0) = 1, \dot{y}(0) = 0.$$

As in section 10.1 we put $wt = \theta, w^{-2} = 1 - \varepsilon \eta(\varepsilon)$, so

$$y'' + y = \varepsilon(\eta y + (1 - \varepsilon \eta)y^2)$$

The aequivalent integral equations are

$$\begin{aligned}
y(\theta) &= \cos \theta + \varepsilon \int_0^\theta \sin(\theta - \tau)(\eta y + (1 - \varepsilon \eta)y^2)d\tau \\
y'(\theta) &= -\sin \theta + \varepsilon \int_0^\theta \cos(\theta - \tau)(\eta y + (1 - \varepsilon \eta)y^2)d\tau
\end{aligned}$$

Condition 10.4 has not been satisfied but we know from a. that for $\varepsilon$ small enough the solutions are periodic. Expanding $y$ and $\eta$ with respect to $\varepsilon$ and applying the periodicity conditions we find $\eta(\varepsilon) = 0.\varepsilon - \frac{5}{6}\varepsilon + \varepsilon^2 \dots$ so, with $T = 2\pi/w, T_0 = 2\pi, T_1 = 0, T_2 = \frac{5}{6}\pi$.

10.2 As all the solutions are bounded as $a \to 0$ it makes sense to rescale $x = ay$; we find

$$
\begin{aligned}
a\ddot{y} + \sin(ay) &= 0 \quad \text{or} \\
\ddot{y} + y &= \tfrac{a^2}{6}y^3 + O(a^4).
\end{aligned}
$$

Applying the Poincaré-Lindstedt method we find

$$
T(a) = 2\pi\left(1 + \frac{a^2}{16} + O(a^4)\right).
$$

10.4 Using section 10.3 we have in 10.17

$$
g = \beta x + (1 - \varepsilon\beta)^{1/2}(1 - x^2)x' + (1 - \varepsilon\beta)h\cos(\theta - \psi)
$$

with expansion

$$
g = \beta a_0 \cos\theta - (1 - a_0^2\cos^2\theta)a_0 \sin\theta + h\cos(\theta - \psi_0) + \varepsilon \ldots
$$

The periodicity conditons 10.18 become to first-order

$$
\begin{aligned}
a_0\beta + h\cos\psi_0 &= 0 \\
-a_0(1 - \tfrac{1}{4}a_0^2) + h\sin\psi_0 &= 0.
\end{aligned}
$$

11.1  a. $(0,0)$ is critical point; the expression

$$
F(x, \dot{x}) = \frac{1}{2}\dot{x}^2 + \frac{1}{2}x^2 + \frac{1}{4}\varepsilon x^4
$$

is a first integral of the equation. Outside $(0,0)$ $F(x, \dot{x}) =$ constant describes closed curves only.

b. Introducing the transformation 11.2-3 yields

$$
\dot{r} = \varepsilon \sin(t + \psi)r^3 \cos^3(t + \psi)
$$

$$
\dot{\psi} = \varepsilon \cos(t + \psi)r^2 \cos^3(t + \psi)
$$

and after averaging

$$
\dot{r}_a = 0, \quad \dot{\psi}_a = \frac{3}{8}\varepsilon r_a^2.
$$

So

$$
\begin{aligned}
x(t) &= r_0 \cos(t - \tfrac{3}{8}\varepsilon r_0^2 t + \psi_0) + O(\varepsilon) \\
\dot{x}(t) &= -r_0 \sin(t - \tfrac{3}{8}\varepsilon r_0^2 t + \psi_0) + O(\varepsilon)
\end{aligned}
$$

on the time-scale $1/\varepsilon$; $r_0$ and $\psi_0$ are determined by the initial conditions.

11.2 Assume that $\phi(t)$ and $\psi(t)$ are two independent solutions of the linear equation. Following section 11.2 we put

$$x = \phi(t)y_1 + \psi(t)y_2$$
$$\dot{x} = \dot{\phi}(t)y_1 + \dot{\psi}(t)y_2$$

to find

$$\dot{y}_1 = \varepsilon\frac{\psi(t)}{\dot{\phi}\psi - \phi\dot{\psi}}f(\cdot,\cdot), \dot{y}_2 = -\varepsilon\frac{\phi(t)}{\dot{\phi}\psi - \phi\dot{\psi}}f(\cdot,\cdot)$$

11.3   a. Putting $x(t) = r(t)\cos(t + \psi(t))$, $\dot{x}(t) = -r(t)\sin(t + \psi(t))$ as in section 11.1 produces

$$\dot{r} = -\varepsilon\sin(t + \psi)f(-r\sin(t + \psi))$$

$$\dot{\psi} = -\frac{\varepsilon}{r}\cos(t + \psi)f(-r\sin(t + \psi)).$$

   b. Choose $r(0) = r_0$, $\psi(0) = \psi_0$ and expand $f(-r\sin(t + \psi)) = \sum_{n=0}^{\infty}f^{(n)}(0)\frac{(-1)^n}{n!}r^n\sin^n(t + \psi)$. Averaging produces

$$\frac{1}{2\pi}\int_0^{2\pi}\sin(t+\psi)f(-r\sin(t+\psi))dt = \sum_{n=0}^{\infty}f^{(2n+1)}(0)\frac{r^{2n+1}}{(2n + 1)!2\pi}\int_0^{2\pi}\sin^{2n+2}t\,dt$$

$$= \sum_{n=0}^{\infty}f^{(2n+1)}(0)\frac{r^{2n+1}(2n + 2)}{2^{2n+2}((n + 1)!)^2}$$

$$\frac{1}{2\pi}\int_0^{2\pi}\cos(t + \psi)f(-r\sin(t + \psi))dt = 0.$$

Solving $r_a$ from

$$\dot{r}_a = -\varepsilon\sum_{n=0}^{\infty}f^{(2n+1)}(0)\frac{r_a^{2n+1}(2n + 2)}{2^{2n+2}((n + 1)!)^2}$$

yields $r(t) = r_a(t) + O(\varepsilon)$, $\psi(t) = \psi_0 + O(\varepsilon)$ on the time-scale $1/\varepsilon$.

   c. The approximation is isochronous but has in general not a constant amplitude.

   d. If $f(\dot{x})$ is even, $r(t) = r_0 + O(\varepsilon)$, $\psi(t) = \psi_0 + O(\varepsilon)$ on the time-scale $1/\varepsilon$. To study the effect of the perturbation one has to obtain higher accuracy than $O(\varepsilon)$ or one has to consider approximations on a longer time-scale than $1/\varepsilon$.

11.4   a. A vector perpendicular to the plane at time $t$ spanned by $\vec{r}$ and $d\vec{r}/dt$ is $\vec{r} \times d\vec{r}/dt$. For its change with time we find

$$\frac{d}{dt}\vec{r} \times \frac{d\vec{r}}{dt} = \frac{d\vec{r}}{dt} \times \frac{d\vec{r}}{dt} + \vec{r} \times \frac{d^2\vec{r}}{dt^2} = 0.$$

d. With $\dot\theta = cr^{-2}e^{-\varepsilon t}$ and taking $c > 0$, $\theta(t)$ is a monotonically increasing function and can be used as a time-like variable.

$$\frac{d\rho}{d\theta} = \frac{d\rho}{dr}\frac{dr}{dt}\Big/\frac{d\theta}{dt} = -\frac{\dot r}{c}e^{\varepsilon t}$$

$$\frac{d^2\rho}{d\theta^2} = \frac{d}{dt}\Big(\frac{d\rho}{d\theta}\Big)\Big/\frac{d\theta}{dt} = -\frac{1}{c}(\ddot r + \varepsilon\dot r)e^{\varepsilon t}\frac{r^2}{c}e^{\varepsilon t}$$

(with the equation for $\ddot r$)

$$= -\frac{1}{r} + \frac{1}{c^2}e^{2\varepsilon t}$$

so

$$\frac{d^2\rho}{d\theta^2} + \rho = \frac{1}{c^2}e^{2\varepsilon t} = u$$

$$\frac{du}{d\theta} = \frac{du}{dt}\Big/\frac{d\theta}{dt} = \frac{2\varepsilon}{c^2}e^{2\varepsilon t}\frac{r^2}{c}e^{\varepsilon t} = \frac{2\varepsilon}{c^3\rho^2}e^{3\varepsilon t} = 2\varepsilon\frac{u^{3/2}}{\rho^2}.$$

e. Differentiation and applying the equations of motion yields

$$\frac{da}{d\theta} = -2\varepsilon\frac{\cos\theta\, u^{3/2}}{(u + a\cos\theta + b\sin\theta)^2}$$

$$\frac{db}{d\theta} = -2\varepsilon\frac{\sin\theta\, u^{3/2}}{(u + a\cos\theta + b\sin\theta)^2}$$

$$\frac{du}{d\theta} = 2\varepsilon\frac{u^{3/2}}{(u + a\cos\theta + b\sin\theta)^2}$$

f.

$$\frac{1}{2\pi}\int_0^{2\pi}\frac{\cos\theta\, u^{3/2}}{(u + a\cos\theta + b\sin\theta)^2}\,d\theta = -\frac{a u^{3/2}}{(u^2 - a^2 - b^2)^{3/2}}$$

$$\frac{1}{2\pi}\int_0^{2\pi}\frac{\sin\theta\, u^{3/2}}{(u + a\cos\theta + b\sin\theta)^2}\,d\theta = -\frac{b u^{3/2}}{(u^2 - a^2 - b^2)^{3/2}}$$

$$\frac{1}{2\pi}\int_0^{2\pi}\frac{u^{3/2}}{(u + a\cos\theta + b\sin\theta)^2}\,d\theta = \frac{u^{5/2}}{(u^2 - a^2 - b^2)^{3/2}}$$

At $\theta = 0$ we have $a(0) = b(0) = 0$, $u(0) = 1/c^2$. For the approximate quantities $\tilde a(\theta), \tilde b\,(\theta), \tilde u(\theta)$ we find $\tilde a(\theta) = \tilde b(\theta) = 0$.

$$\frac{d\tilde u}{d\theta} = 2\varepsilon\tilde u^{-1/2}$$

or

$$\tilde u(\theta) = \Big(\frac{1}{c^3} + 3\varepsilon\theta\Big)^{2/3}$$

The approximations have accuracy $O(\varepsilon)$ on the $1/\varepsilon$- time-scale $\theta$; as long as $r$ is bounded away from zero the same time-scale holds in $t$. Also, with these initial conditions,

$$\rho(\theta) = \tilde u(\theta) + O(\varepsilon) \text{ on } 1/\varepsilon.$$

As
$$u = \frac{1}{c^2}e^{2\varepsilon t} \quad \text{we have} \quad \frac{1}{c^2}e^{2\varepsilon t} = (\frac{1}{c^3} + 3\varepsilon\theta)^{2/3} \quad \text{or}$$

$$\theta(t) = \frac{1}{3\varepsilon c^3}(e^{3\varepsilon t} - 1)$$

As we approximate $\rho = \frac{1}{r} = \tilde{u}$ we have the approximation

$$\tilde{r}(t) = c^2 \varepsilon^{-2\varepsilon t}.$$

**11.5** The averaged equation is

$$\dot{y} = \varepsilon(a - y)$$

with $y = a$ an attracting critical point. So according to theorem 11.5 there exists a $2\pi$-periodic solution $\phi(t,\varepsilon)$ of the equation for $x$ with the property $\lim_{\varepsilon \to 0}\phi(t,\varepsilon) = a$. According to theorem 11.6 the periodic solution is asymptotically stable.

Note that the solutions of the equation for $x$ are given by

$$x(t) = a - \frac{\varepsilon}{1 + \varepsilon^2}\cos t + \frac{\varepsilon^2}{1 + \varepsilon^2}\sin t + ce^{-\varepsilon t}$$

with $c$ an arbitrary constant.

**11.6** The averaged system is $\dot{y} = \varepsilon g(y)$ with

$$g(y) = \begin{matrix} y_1(y_2 + y_3 - 2) \\ y_2(y_3 + y_1 - 2) \\ y_3(y_1 + y_2 - 2) \end{matrix}$$

Apart from the critical point $y = (0,0,0)$ which is asymptotically stable and which is also a critical point of the original equation we have 4 critical points of the averaged system:
$(0,2,2), (2,0,2), (2,2,0), (1,1,1)$.
Applying theorems 11.5-6 we find near each of these 4 critical points existence of a $2\pi$-periodic solution which is unstable.

**11.7** From the discussion of example 11.9 we know that to prove existence we have to reduce the order of the equation. Putting $x = r\cos(t + \psi)$, $\dot{x} = -r\sin(t + \psi)$, $t + \psi = \theta$ we find

$$\frac{dr}{d\theta} = \frac{\dot{r}}{1 + \dot{\psi}} = \varepsilon\frac{r\sin^2\theta(1 - ar^2\cos^2\theta - br^2\sin^2\theta)}{1 + \varepsilon\ldots}$$

Averaging of the $O(\varepsilon)$ terms over $\theta$ produces

$$\frac{dr_a}{d\theta} = \frac{1}{2}\varepsilon r_a(1 - \frac{1}{4}ar_a^2 - \frac{3}{4}br_a^2)$$

with critical point $p = 2/\sqrt{a + 3b}$. Condition 11.50 has been satisfied so a $2\pi$-periodic solution $r(\theta)$ exists (theorem 11.5) and as the original equation is autonomous, this corresponds with a time-periodic solution. The eigenvalue of the critical point is negative so the periodic solution is asymptotically stable.

To obtain an asymptotic approximation we still have to average the equation for $\psi$:

$$\dot{\psi} = \varepsilon \cos(t + \psi) \sin(t + \psi)(1 - ar^2 \cos^2(t + \psi) - br^2 \sin^2(t + \psi)),$$

which produces zero. It is no restriction of generality to put $\psi(0) = 0$ so the periodic solution is approximated by

$$\frac{2}{\sqrt{a + 3b}} \cos t.$$

11.8    a. Using the solutions in the case $\varepsilon = 0$ we have the variation of constants transformation $x, y \rightarrow u, v$

$$x = \frac{u e^t}{1 - u + u e^t} \quad , y = \frac{v e^t}{1 - v + v e^t}$$

we find after substitution in the equations

$$\dot{u} = -\varepsilon u v \frac{1 - u + u e^t}{1 - v + v e^t} e^t \quad , u(0) = x_0$$
$$\dot{v} = -\varepsilon u v \frac{1 - v + v e^t}{1 - u + u e^t} e^t \quad , v(0) = y_0.$$

b. $\tau = e^t$ yields

$$\frac{du}{d\tau} = -\varepsilon u v \frac{1 - u + u\tau}{1 - v + v\tau} \quad , u(\tau = 1) = x_0$$
$$\frac{dv}{d\tau} = -\varepsilon u v \frac{1 - v + v\tau}{1 - u + u\tau} \quad , v(\tau = 1) = y_0.$$

As $\displaystyle\lim_{T \to \infty} \frac{1}{T} \int_1^T \frac{1 - p + p\tau}{1 - q + q\tau} d\tau = \frac{p}{q}$ the averaged system in the sense of theorem 11.3 is

$$\frac{du_a}{d\tau} = -\varepsilon u_a^2 \, , \, \frac{dv_a}{d\tau} = -\varepsilon v_a^2.$$

So $u_a(\tau) = \frac{1}{x_0^{-1} - \varepsilon + \varepsilon\tau}$ , $v_a(\tau) = \frac{1}{y_0^{-1} - \varepsilon + \varepsilon\tau}$. Transforming $\tau = e^t$ and $u, v \rightarrow x, y$ produces the approximations

$$x_a(t) = \frac{x_0 e^t}{1 + x_0(1 + \varepsilon)(e^t - 1)}, y_a(t) = \frac{y_0 e^t}{1 + y_0(1 + \varepsilon)(e^t - 1)}.$$

c. The approximation holds in $\tau$ on the time-scale $1/\varepsilon$, in the original time $t$ on the time-scale $ln(1/\varepsilon)$. According to theorem 11.3 the error is given by

$$\delta(\varepsilon) = \sup_{u,v \in D} \sup_{0 \leq \tau \leq C} \varepsilon \| \int_1^\tau (\frac{1-u+u\tau}{1-v+v\tau} - \frac{u}{v})d\tau \| = O(\varepsilon \mid ln\varepsilon \mid).$$

On the other hand we know that $(\frac{1}{1+\varepsilon}, \frac{1}{1+\varepsilon})$ is an attracting node (exercise 7.4) and that $(x_a(ln(1/\varepsilon)), y_a(ln(1/\varepsilon))) = (1+O(\varepsilon), 1+O(\varepsilon))$. So on the time-scale of validity the approximate solution reaches an $O(\varepsilon)$ neighbourhood of the attracting node. According to the Poincaré-Lyapunov theorem 7.1 there exists an $\varepsilon_0$ such that for $0 \leq \varepsilon \leq \varepsilon_0$ the solution will not leave this neighbourhood. As $\varepsilon = O(\varepsilon \mid ln\varepsilon \mid)$ (even $o(\varepsilon \mid ln\varepsilon \mid)$) we have $(x(t), y(t)) = (x_a(t), y_a(t)) + O(\varepsilon \mid ln\varepsilon \mid)$ for all time.

11.9   a. A suitable transformation is $\rho, d\rho/d\theta \to a, \psi$ by

$$\rho = \mu + a\cos(\theta + \psi)$$
$$\frac{d\rho}{d\theta} = -a\sin(\theta + \psi)$$

producing

$$\dot{a} = -\varepsilon \sin(\theta + \psi)(\mu + a\cos(\theta + \psi))^2$$
$$\dot{\psi} = -\frac{\varepsilon}{a}\cos(\theta + \psi)(\mu + a\cos(\theta + \psi))^2$$

b. Averaging produces zero for the first equation and $\dot{\psi}_a = -\varepsilon\mu$ for the second. Applying the initial conditions $a(0) = a_0, \psi(0) = \psi_0$ we find

$$\rho = \mu + a_0 \cos(\theta - \varepsilon\mu\theta + \psi_0) + O(\varepsilon)$$
$$\frac{d\rho}{d\theta} = -a_0 \sin(\theta - \varepsilon\mu\theta + \psi_0) + O(\varepsilon)$$

on the time-scale $1/\varepsilon$.
The amount $\varepsilon\mu\theta$ will advance the perihelium (or periastron) of the orbit. In the solar system this relativistic perturbation is most prominent in the case of Mercury.

11.10   a. There are several possibilities; one runs as follows:

$$\ddot{y} = -4\dot{x} = -4y - 4\varepsilon(\frac{1}{16}\dot{y}^2 \sin 2t - \sin 2t).$$

Transformation 11.15 becomes $y(t) = u(t)\cos 2t + \frac{1}{2}v(t)\sin 2t$, $\dot{y}(t) = -2u(t)\sin 2t + v(t)\cos 2t$. Lagrange standard form

$$\dot{u} = 2\varepsilon \sin 2t[\frac{1}{16}(-2u\sin 2t + v\cos 2t)^2 \sin 2t - \sin 2t]$$
$$\dot{v} = -4\varepsilon \cos 2t[\frac{1}{16}(-2u\sin 2t + v\cos 2t)^2 \sin 2t - \sin 2t]$$

Averaging produces

$$\dot{u}_a = \varepsilon\left(\tfrac{3}{16}u_a^2 + \tfrac{1}{64}v_a^2 - 1\right)$$
$$\dot{v}_a = \tfrac{1}{8}\varepsilon u_a v_a$$

b. From theorem 11.5-6 we find $(u_a, v_a)$ correspond with periodic solutions if $(u_a, v_a) = (0, \pm 8)$ (unstable), $(\tfrac{4}{3}\sqrt{3}, 0)$ (unstable) and $(-\tfrac{4}{3}\sqrt{3}, 0)$ (stable).

11.11  a.

$$\dot{r} = -\varepsilon\tfrac{r^3}{\omega}\sin^2\phi(\cos^2\phi - \omega^2\sin^2\phi) - \tfrac{1}{2}\varepsilon r\sin^2\phi$$
$$\dot{\phi} = \omega - \tfrac{1}{2}\varepsilon\sin\phi\cos\phi - \tfrac{\varepsilon}{\omega}r^2\sin\phi\cos\phi(\cos^2\phi - \omega^2\sin^2\phi)$$
$$\dot{\omega} = \varepsilon r^2(\cos^2\phi - \omega^2\sin^2\phi)$$

b. Averaging over $\phi$ produces

$$\dot{r}_a = -\varepsilon\tfrac{r_a^3}{\omega_a}\tfrac{1}{8} + \varepsilon r_a^3\omega_a\tfrac{3}{8} - \tfrac{1}{4}\varepsilon r_a$$
$$\dot{\omega}_a = \varepsilon r_a^2\tfrac{1}{2}(1 - \omega_a^2)$$
$$r_a(0) = r(0), \omega_a(0) = \omega(0).$$

According to theorem 11.4 we have $r(t) - r_a(t)$ and $\omega(t) - \omega_a(t) = O(\varepsilon)$ on the time-scale $1/\varepsilon$.

c. Consider $\dot{x} = \varepsilon X(\phi, x), \dot{\phi} = \Omega(x)$ with $\phi \in S^1$ and averaged equation $\dot{y} = \varepsilon X^0(y)$. Replacing $t$ by the time-like variable $\phi$ produces the standard form

$$\dot{x} = \varepsilon\frac{X(\phi, x)}{\Omega(x)}, \text{ averaged equation } \dot{y} = \varepsilon\frac{X^0(y)}{\Omega(y)}.$$

In calculating the determinant condition 11.50 we find for the critical point $P$ with $X^0(y)\mid_P = 0$ that a solution periodic in $\phi$ exists if

$$|\partial X^0(y)/\partial y\mid_{y=P} \neq 0.$$

d. Critical point $r_a = 1, \omega_a = 1$. The Jacobi determinant does-not vanish so a periodic solution exists.

13.1 $(0,0)$ has eigenvalues $-1$ and $0$; the centre manifold is one-dimensional and is represented by

$$y = h(x) = ax^2 + bx^3 + \ldots$$

Substitution in the equation for $y$ produces

$$\frac{dh}{dx}\dot{x} = \frac{dh}{dx}(xh(x)) = -h(x) + 3x^2$$

or

$$(2ax + 3bx^2 + \ldots)(ax^3 + bx^4 + \ldots) = -ax^2 - bx^3 + \ldots + 3x^2$$

We find $a = 3$, $b = 0$ so that $h(x) = 3x^2 + O(x^4)$; the flow in the centre manifold is determined by

$$\dot{x} = xh(x) = 3x^3 + O(x^5)$$

so $(0,0)$ is unstable.

13.2 We find a Taylor expansion with coefficients zero; this could be expected as the centre manifold is non-analytic.

13.3 The eigenvalues are $\pm i$ and $-1$, so the centre manifold is two-dimensional. $E_s$ coincides with the $z$-axis, $E_c$ coincides with the $x, y$-plane. Approximate the centre manifold by

$$z = h(x,y) = ax^2 + 2bxy + cy^2 + \ldots$$

so that

$$\dot{z} = \frac{\partial h}{\partial x}\dot{x} + \frac{\partial h}{\partial y}\dot{y} = -z - (x^2 + y^2) + z^2 + \sin x^3.$$

Substitution of the expansion and equating the coefficients of equal powers produces $a = -1$, $b = 0$, $c = -1$. $W_c$ is approximated by $z = -x^2 - y^2 + \ldots$
Substitution of $z = h(x,y)$ in the equations for $x$ and $y$ yields

$$\dot{x} = -y - x^3 - xy^2 + \ldots$$
$$\dot{y} = x - x^2y - y^3 + \ldots$$

Using polar coordinates one finds that $(0,0,0)$ is stable.

13.4 The equilibrium solutions are $1 + \mu \pm (2\mu + \mu^2)^{1/2}$. The solutions with minus sign are stable, with plus sign unstable for $\mu > 0$, $\mu < -2$; at $\mu = 0, -2$ the solutions are unstable.

13.5 Writing down the homology equation for $h(y)$ we note that the equation is of first order and, when written in explicit form, its righthand side has a convergent Taylor series with respect to $h$ and $y$ in a neighbourhood of $h = y = 0$. Note that as $\lambda \neq 0$, there is no resonance. It follows from the elementary theory of differential equations that $h(y)$ can be expanded in a convergent power series near $y = 0$.

13.6 Saddles and centres.

**13.7** The Brusselator has one critical point: $x = a$, $y = b/a$. Linearising near this critical point leads to the eigenvalue equation $\lambda^2 + (a^2 - b + 1)\lambda + a^2 = 0$ Hopf bifurcation is possible if in $a, b$-parameter space one crosses the curve $a^2 = b - 1$. If for instance $a = 1$, one has a limitcycle for $b > 2$, no limit cycle for $b < 2$ (supercritical bifurcation).

**13.8**    a. Substitution produces

$$\dot{x} = \dot{u} + \varepsilon(2a_1 u\dot{u} + a_2\dot{u}v + a_2 u\dot{v} + 2a_3 v\dot{v}) + \varepsilon^2 \ldots =$$

$$v + \varepsilon(b_1 u^2 + b_2 uv + b_3 v^2) + \varepsilon^2 \ldots$$

$$\dot{y} = \dot{v} + \varepsilon(2b_1 u\dot{u} + b_2\dot{u}v + b_2 u\dot{v} + 2b_3 v\dot{v}) + \varepsilon^2 \ldots =$$

$$-u - \varepsilon(a_1 u^2 + a_2 uv + a_3 v^2) + \varepsilon u^2 + \varepsilon^2 \ldots$$

As $\dot{u} = v + \varepsilon \ldots$ and $\dot{v} = -u + \varepsilon \ldots$ this reduces to

$$\dot{u} + \varepsilon(2a_1 uv + a_2 v^2 - a_2 u^2 - 2a_3 uv) + \varepsilon^2 \ldots =$$

$$v + \varepsilon(b_1 u^2 + b_2 uv + b_3 v^2) + \varepsilon^2 \ldots$$

$$\dot{v} + \varepsilon(2b_1 uv + b_2 v^2 - b_2 u^2 - 2b_3 uv) + \varepsilon^2 \ldots =$$

$$-u - \varepsilon(a_1 u^2 + a_2 uv + a_3 v^2) + \varepsilon u^2 + \varepsilon^2 \ldots$$

The equations for $u$ and $v$ donot contain quadratic terms if $a_1 = \frac{1}{3}, a_2 = 0, a_3 = \frac{2}{3}, b_1 = 0, b_2 = -\frac{2}{3}, b_3 = 0$. We have

$$\dot{u} = v + \varepsilon^2 \,, \quad \dot{v} = -u + \varepsilon^2 \ldots$$

b. Transforming $u = r\cos(t + \psi), v = -r\sin(t + \psi)$ yields the standard form

$$\dot{r} = \varepsilon^2 \ldots \,, \quad \dot{\psi} = \varepsilon^2 \ldots$$

One can apply for instance theorem 9.1 putting $\tau = \varepsilon t$ ($f_0$ in the theorem vanishes). The result is the same $O(\varepsilon)$ approximation on the time-scale $1/\varepsilon$ as by the averaging method.

**13.9**    a. The eigenvalue equation is $\lambda^2 + (1 - a^2)\lambda - (2a^2 + 6a) = 0$ with solutions $\lambda_1$ and $\lambda_2$; $\lambda_1 + \lambda_2 = -1 + a^2$, $\lambda_1\lambda_2 = -2a(a + 3)$. We have asymptotic stability by theorem 7.1 if $-1 < a < 0$.

b. Denoting the righthand side of the system by the 2-vector $F(x,y)$ we have the possibility of bifurcation if

$$\frac{\partial F(x,y)}{\partial(x,y)}(0,0) = 0$$

which is satisfied for $a = -3$ and $a = 0$. To illustrate this take $a$ near zero, $f(x,y) = x^2$, $g(x,y) = -y^2$. If $a = 0$ there is one solution $((0,0))$, if $a$ increases 4 critical points branch off (intersections of an ellips and a hyperbole in the $x,y$-plane).

c. Eigenvalues with real part zero exist if $a = -3, -1, 0$.

d. $a = -1$.

13.10 Linear analysis yields eigenvalues $-2$ and $0$. The $y$-axis is centre manifold (one can also obtain this result by carrying out the series expansion of section 13.4). In the centre manifold the flow is attracting so according to theorem 13.4 $(0,0)$ is stable.

15.1 According to theorem 2.1 of Liouville, the phase-flow is volume-preserving. So the trace of the linearisation matrix has to be zero, the eigenvalues have to be symmetric around zero in the complex plane. The constants $a$ and $b$ are real, then we have

 a. $\pm ai, \pm bi \quad \pm a, \pm bi \quad \pm a, \pm b \quad a \pm bi, -a \pm bi$ (4 cases).

 b. For periodic solutions we interprete the question as follows. One characteristic exponent is zero because of the autonomous character of the equations, one is zero as we have a family of periodic solutions parameterised by the energy. A Poincaré-mapping of a neighbourhood of the periodic solution for a fixed value of the energy is two-dimensional and area-preserving with the periodic solution showing up as a fixed point. The possible eigenvalues are $\pm ai$ and $\pm a$ (2 cases).

 c. If $n = 3$ we have, as in b, two characteristic exponents zero; the Poincaré-mapping for a fixed value of the energy is four-dimensional and volume-preserving. The possible eigenvalues are the same as in a (4 cases).

15.2 a. The equations of motion are $\dot{q}_1 = \omega_1 p_1 + \varepsilon \ldots$, $\dot{p}_1 = -\omega_1 q_1 + \varepsilon \ldots$, $\dot{q}_2 = \omega_2 p_2 + \varepsilon \ldots$, $\dot{p}_2 = -\omega_2 q_2 + \varepsilon \ldots$. The eigenvalues of the linearised system are $\pm \omega_1 i$ and $\pm \omega_2 i$. We have resonance as in example 13.5.

 b. The resonance relation, section 13.2, becomes

$$\pm \omega_i = (m_1 - m_2)\omega_1 + (m_3 - m_4)\omega_2 \,, i = 1, 2$$

$m_1, \ldots, m_4 \in \{0\} \cup \mathbb{N}$ and $m_1 + \ldots + m_4 \geq 2$.
First-order resonance arises if $m_1 + \ldots + m_4 = 2$, $m_1 + m_2 > 0$ and $m_3 + m_4 > 0$. This produces $\omega_1 = 2\omega_2$ or $2\omega_1 = \omega_2$.

 c. For second-order resonance we have $m_1 + \ldots + m_4 = 3$, $m_1 + m_2 > 0$ and $m_3 + m_4 > 0$. This produces $\omega_1 = \omega_2$, $\omega_1 = 3\omega_2$ or $3\omega_1 = \omega_2$.

d. In the case of first-order resonance the normal form contains nontrivial terms $O(\varepsilon)$ so the time-scale by averaging is $1/\varepsilon$. For second-order resonance the $O(\varepsilon)$ terms are removed, the time-scale of validity becomes $1/\varepsilon^2$.

15.3    a. Equations of motion:

$$\ddot{q}_1 + 4q_1 = -4c\varepsilon q_1 + \varepsilon(aq_1^2 + bq_2^2)$$
$$\ddot{q}_2 + q_2 = \varepsilon 2bq_1 q_2$$

Transformation (15.8) produces

$$\dot{r}_1 = c\varepsilon r_1 \sin(4t + 2\phi_1) - \tfrac{1}{2}\varepsilon \sin(2t + \phi_1)(ar_1^2 \cos^2(2t + \phi_1) + br_2^2 \cos^2(t + \phi_2))$$
$$\dot{\phi}_1 = c\varepsilon + c\varepsilon \cos(4t + 2\phi_1) - \tfrac{\varepsilon}{2r_1} \cos(2t + \phi_1) \times (ar_1^2 \cos^2(2t + \phi_1) + br_2^2 \cos^2(t + \phi_2))$$
$$\dot{r}_2 = -\varepsilon 2br_1 r_2 \sin(t + \phi_2) \cos(t + \phi_2) \cos(2t + \phi_1)$$
$$\dot{\phi}_2 = -\varepsilon 2br_1 \cos(2t + \phi_1) \cos^2(t + \phi_2)$$

b. The averaged system is

$$\dot{\rho}_1 = -\varepsilon \tfrac{b}{8}\rho_2^2 \sin(\psi_1 - 2\psi_2)$$
$$\dot{\psi}_1 = c\varepsilon - \varepsilon \tfrac{b}{8}\tfrac{\rho_2^2}{\rho_1} \cos(\psi_1 - 2\psi_2)$$
$$\dot{\rho}_2 = \varepsilon \tfrac{1}{2}b\rho_1 \rho_2 \sin(\psi_1 - 2\psi_2)$$
$$\dot{\psi}_2 = -\varepsilon \tfrac{1}{2}b\rho_1 \cos(\psi_1 - 2\psi_2)$$

If $c = 0$ the system is of the form (15.9), $b$ can then be scaled away by $t \to \tau := bt$.
If $b = 0$ their is no interaction; note that in this case the original system is already decoupled and integrable.

c. The integrals of the averaged system are

$$4\rho_1^2 + \rho_2^2 = 2E_0 \text{ (cf.15.10)}$$

$$\frac{1}{2}b\rho_1 \rho_2^2 \cos(\psi_1 - 2\psi_2) - 2c\rho_1^2 = I \text{ (cf.15.11)}$$

with $E_0$ and $I$ constants.

d. $q_2(t) = 0$, $t \in \mathbb{R}$ produces an exact normal mode solution of the original system; it is recovered in the averaged system by putting $\rho_2 = 0$.
As in section 15.2 we obtain other periodic solutions by requiring that $\psi_1 - 2\psi_2 = 0$ (in-phase) or $\pi$ (out-phase) and $\dot{\rho}_1 = \dot{\rho}_2 = 0$.

Existence in-phase solution: $c > -\frac{1}{2}b\sqrt{2E_0}$.
Existence out-phase solution: $c < \frac{1}{2}b\sqrt{2E_0}$.
At $c = \pm\frac{1}{2}b\sqrt{2E_0}$ we have a bifurcation. For instance considering the Poincaré map for $b = 1, c = 0$ in fig. 15.4, we increase c. The two fixed points (centres) corresponding with the out-phase solution move towards the saddle point at $(0,0)$ (corresponding with the normal mode). At $c = \frac{1}{2}b\sqrt{2E_0}$ the three critical points (2 periodic solutions) coalesce, resulting in a centre for $c > \frac{1}{2}b\sqrt{2E_0}$.

e. Analysing the critical points in the Poincaré map we find that the in-phase and out-phase periodic solutions are stable. The normal mode is unstable if $-\frac{1}{2}b\sqrt{2E_0} \leq c \leq \frac{1}{2}b\sqrt{2E_0}$ (if $b > 0$; if $b < 0$ reverse the end-points). Outside this parameter interval the normal mode is stable.

15.4   a. The equations can be derived from the Hamiltonian $H = \frac{1}{2}(q_1^2 + p_1^2) + \frac{1}{2}(q_2^2 + p_2^2) - (\frac{1}{3}aq_1^3 + bq_1q_2^2)$, which is positive definite near the origin of phase-space (cf. section 15.1).

b. Putting $q_2(t) = 0, t \in \mathbb{R}$ the second equation is satisfied. The first equation becomes $\ddot{q}_1 + q_1 = aq_1^2$, see example 2.11 where $a = \frac{1}{2}$. If $a = 0$, these solutions are periodic for all values of the energy; if $a \neq 0$ there is a critical point at $(q_1, \dot{q}_1, q_2, \dot{q}_2) = (1/a, 0, 0, 0)$, which is unstable. If $a \neq 0$ these particular solutions are periodic if $0 < h < 1/6a^2$. Putting $q_1 = \lambda q_2$ the equations become

$$\ddot{q}_2 + q_2 = (a\lambda + b/\lambda)q_2^2$$
$$\ddot{q}_2 + q_2 = 2b\lambda q_2^2$$

The equations are consistent if $\lambda^2 = b/(2b-a)$ which requires the righthand side to be positive.

15.5   a. Equations of motion $\ddot{q}_1 + q_1 = q_1^2 - q_2^2$, $\ddot{q}_2 + q_2 = -2q_1q_2$ produce critical points $(0,0,0,0)$ centre-centre, $(1,0,0,0)$ and $(-\frac{1}{2}, 0, \pm\frac{1}{2}\sqrt{3}, 0)$. The first is stable, the other three unstable.

b. Near $(0,0,0,0)$ the energy-surface is compact, it breaks up at the energy level $H = 1/6$ where the three unstable critical points are located. Compactness for $0 \leq H < 1/6$.

c. We apply the results of exercise 15.4. The $(q_1, p_1)$-normal mode exists as a periodic solution for $0 < H < 1/6$. Furthermore $\lambda^2 = 1/3$, producing an in-phase periodic solution given by $q_1 = \frac{1}{3}\overline{3}q_2$, $\ddot{q}_2 + q_2 = -\frac{2}{3}\overline{3}q_2^2$ and an out- phase periodic solution given by $q_1 = -\frac{1}{3}\sqrt{3}q_2$, $\ddot{q}_2 + q_2 = \frac{2}{3}\sqrt{3}q_2^2$. In both cases we find closed curves, i.e. periodicity if $0 < H < 1/6$.

d. No. See exercise 15.3c where for a fixed value of $c$, periodic solutions branch off the normal mode at a certain energy level.

15.6   a. (Solution by I. Hoveijn).
We put $\frac{dx}{d\tau} = \frac{\partial F}{\partial y}, \frac{dy}{d\tau} = -\frac{\partial F}{\partial x}$; then we have

$$\frac{dx}{d\tau} = \frac{\dot{x}}{\dot{\tau}} = \frac{f(x,y)}{\dot{\tau}} = \frac{\partial F}{\partial y} \quad \text{and} \quad \frac{dy}{d\tau} = \frac{\dot{y}}{\dot{\tau}} = \frac{g(x,y)}{\dot{\tau}} = -\frac{\partial F}{\partial x}$$

with the requirement

$$\dot{\tau} = \frac{f(x,y)}{\partial F/\partial y} = \frac{g(x,y)}{-\partial F/\partial x}.$$

This has been satisfied as $L_t F = \frac{\partial F}{\partial x}\dot{x} + \frac{\partial F}{\partial y}\dot{y} = \frac{\partial F}{\partial x}f(x,y) + \frac{\partial F}{\partial y}g(x,y) = 0$.

b. The transformation is formal as $\tau$ may only be locally time-like.

c. We have $\dot{x} = ax - bxy$, $\dot{y} = bxy - cy$, $F(x,y) = bx - c\ln x + by - a\ln y$. Introduce $\dot{\tau} = -xy$ so that

$$\frac{dx}{d\tau} = b - \frac{a}{y} = \frac{\partial F}{\partial y}, \frac{dy}{d\tau} = \frac{c}{x} - b = -\frac{\partial F}{\partial x}.$$

Note that for $x > 0, y > 0, \tau$ is time-like.

d. No as we have three equations to be satisfied for $\dot{\tau}$ and one integral only.

References

**Abraham** , R.H., Shaw, C.B., *Dynamics - the geometry of behaviour (4 vols)*, Aerial Press, Santa Cruz, CA, 1982-1988.

**Amann,** H.: *Gewöhnliche Differentialgleichungen*, De Gruyter, Berlin 1983.

**Andersen** , G.M., Geer, J.F.: *Power series expansions for the frequency and period of the limit cycle of the van der Pol equation*, SIAM J. Appl. Math. **42**, 678-693, 1982.

**Arnold** , V.I., *Mathematical Methods of Classical Mechanics*, Springer-Verlag, New York, 1978 (Russian edition, Moscow 1974, French translation, Mir, Moscow 1976).

**Arnold** , V.I., *Ordinary differential equations*, MIT Press, 1978.

**Arnold** , V.I., *Geometrical methods in the theory of ordinary differential equations*, Springer-Verlag, New York, 1983 (Russian edition, Moscow 1978, French translation, Mir, Moscow 1980).

**Arnold** , V.I., Kozlov, V.V., Neihstadt, A.I., *Dynamical Systems III*, Encyclopaedia of Math. Sciences, (V.I. Arnold, ed.) vol.3, Springer-Verlag, Berlin Heidelberg, 1988.

**Bergé** , P. Pomeau, Y., Vidal, Ch., *L'Ordre dans le chaos*, Hermann, Paris, 1984 (English translation: Order within Chaos, Wiley, Chichester, 1987).

**Bhatia** , N.P. en Szegö, G.P.: *Stability theory of dynamical systems*, Springer-Verlag, Berlin Heidelberg (Grundlehren 161), 1970.

**Bogoliubov** , N.N. and Mitropolsky, Yu.A., *Asymptotic methods in the theory of nonlinear oscillations*, Gordon and Breach, New York, 1961.

**Carr** , J. *Applications of centre manifold theory*, Appl. Math. Sciences 35, Springer-Verlag, New York, 1981.

**Cesari** , L.: *Asymptotic behaviour and stability problems in ordinary differential equations*, Springer-Verlag, Berlin Heidelberg, 1971.


**Coddington** , E.A. and Levinson, N.: *Theory of ordinary differential equations*, Mc-Graw-Hill Book Co., New York, 1955.


**Devaney** , R.L.: *An introduction to Chaotic Dynamical Systems*, Benjamin/Cummings, Menlo Park, CA, 1986.


**Dijksterhuis** , E.J.: *De mechanisering van het wereldbeeld*, J.M. Meulenhoff, Amsterdam, 1950; English translation: The mechanization of the world picture, Galaxy Books, London etc., 1969.


**Eckhaus** , W., *Asymptotic Analysis of Singular Perturbations*, North-Holland, Amsterdam, 1979.


**Grasman** , J., *Asymptotic methods for relaxation oscillations and applications*, Springer-Verlag, New York, Appl. Math. Sciences 63, 1987.


**Grauert** , H. en Fritzsche, K.: *Several complex variables*, Springer-Verlag, New York 1976.


**Guckenheimer** , J. and Holmes, P., *Nonlinear oscillations, dynamical systems and bifurcations of vector fields*, Appl. Math. Sciences 42, Springer-Verlag, New York, 1983.


**Hahn** , W.: *Stability of motion*, Springer-Verlag, Berlin Heidelberg, 1967.


**Hale** , J.K.: *Ordinary differential equations*, Wiley-Interscience, New York, 1969.


**Hartman** , P.: *Ordinary differential equations*, Wiley, New York, 1964 ($2^d$ ed. Birkhäuser, Boston Basel Stuttgart, 1982).

**Jakubovič** , V.A. and Staržinskij, V.M.: *Linear differential equations with periodic coefficients*, Wiley, New York, 1975.

**Kevorkian** , J., *Perturbation techniques for oscillatory systems with slowly varying coefficients*, SIAM Review 29, pp.391-461, 1987.

**Knobloch** , H.W. and Kappel, F.: *Gewöhnliche Differentialgleichungen*, B.G. Teubner, Stuttgart, 1974.

**Lagrange** , J.L., *Mécanique Analytique* (2 vols.) 1788 (édition Albert Blanchard, Paris, 1965).

**Lloyd** , N.G., *Limit cycles of Polynomial Systems - Some Recent Developments*, London Math. Soc. Lect. Note Ser. 127, pp.192-234, Cambridge University Press, 1987.

**Magnus** , W. and Winkler, S.: *Hill's equation*, Interscience, New York, 1966.

**Marsden** , J.E. and McCracken, M.F., *The Hopf bifurcation and its applications*, Appl. Math. Sciences 19, Springer-Verlag, New York 1976.

**Milnor** , J.: *Morse Theory*, Princeton University Press, Princeton 1963.

**Nayfeh** , A.H. and Mook, D.T., *Nonlinear Oscillations*, Wiley-Interscience, New York, 1979.

**Nusse** , H.E., *Complicated dynamical behaviour in discrete population models*, Nieuw Archief Wiskunde (4) 2, pp. 43-81, 1984.

**Reyn** , J.W., *Phase portraits of a quadratic system of differential equations occurring frequently in applications*, Nieuw Archief Wiskunde, (4) 5 pp. 107-115, 1987.

**Roseau** , M.: *Vibrations non linéaires et théorie de la stabilité*, Springer-Verlag, Berlin Heidelberg, 1966.

**Sanders** , J.A. and Verhulst, F., *Averaging methods in nonlinear dynamical systems*, Appl. Math. Sciences 59, Springer-Verlag, New York, 1985.

**Sansone** , G. and Conti, R.: *Nonlinear differential equations*, Pergamon Press, New York, 1964.

**Schuster** , H., Deterministic Chaos, Physik Verlag, Weinheim, 1984.

**Sparrow** , C., *The Lorenz equations: bifurcations, chaos and strange attractors*, Appl. Math. Sciences 41, Springer-Verlag, New York, 1982.

**Van den Broek** , B and Verhulst, F., *Averaging techniques and the oscillator-flywheel problem*, Nieuw Archief Wiskunde (4) 5, pp.185-206, 1987.

**Verhulst** , F., *Asymptotic analysis of Hamiltonian systems*, in Asymptotic Analysis II, Lecture Notes Math. 985, pp.137-193, Springer-Verlag, 1983.

**Walter** , W.: *Gewöhnliche Differentialgleichungen*, Springer-Verlag, Berlin, 1976.

**Weinstein** , A.: *Normal modes for nonlinear Hamiltonian systems*, Inv. Math. 20, pp. 47-57, 1973.

**Ye Yan - Qian** et al., *Theory of limit cycles*, American Math. Soc. Translation Math. Monographs 66, 1986.

# Index

# Encyclopaedia of Mathematical Sciences

**Editor-in-chief: R. V. Gamkrelidze**

## Springer-Verlag – *synonymous with quality in publishing*

- Our **Encyclopaedia of Mathematical Sciences** is much more than just a dictionary. It gives complete and representative coverage of relevant contemporary knowledge in mathematics.
- The volumes are monographs in themselves and contain the principal ideas of the underlying proofs.
- The authors and editors are all distinguished researchers.
- The average volume price makes **EMS** an affordable acquisition for both personal and professional libraries.

## Dynamical Systems

Volume 1: **D. V. Anosov, V. I. Arnold** (Eds.)

### Dynamical Systems I

**Ordinary Differential Equations and Smooth Dynamical Systems**
1988. IX, 233 pp. ISBN 3-540-17000-6

Volume 2: **Ya. G. Sinai** (Ed.)

### Dynamical Systems II

**Ergodic Theory with Applications to Dynamical Systems and Statistical Mechanics**
1989. IX, 281 pp. 25 figs. ISBN 3-540-17001-4

Volume 3: **V. I. Arnold** (Ed.)

### Dynamical Systems III

1988. XIV, 291 pp. 81 figs. ISBN 3-540-17002-2

Volume 4: **V. I. Arnold, S. P. Novikov** (Eds.)

### Dynamical Systems IV

**Symplectic Geometry and its Applications**
1989. VII, 283 pp. 62 figs. ISBN 3-540-17003-0

Volume 5: **V. I. Arnold** (Ed.)

### Dynamical Systems V

**Theory of Birfurcations and Catastrophes**
1990. Approx. 280 pp. ISBN 3-540-18173-3

Springer-Verlag
Berlin Heidelberg New York
London Paris Tokyo
Hong Kong

Volume 6: **V. I. Arnold** (Ed.)

### Dynamical Systems VI

1990. ISBN 3-540-50583-0

Springer

# Applied Mathematical Sciences

**Editors: F. John, J. E. Marsden, L. Sirovich**

Volume 63

**J. Grasman,** State University of Utrecht, The Netherlands

# Asymptotic Methods for Relaxation Oscillations and Applications

1987. XIII, 321 pp. 89 figs. ISBN 3-540-96513-0

**Contents:** Introduction. – Free oscillation. – Forced oscillation and mutual entrainment. – The Van der Pol oscillator with a sinusoidal forcing term. – Appendices. – Literature. – Author Index. – Subject Index.

Volume 72

**P. Lochak,** Paris; **C. Meunier,** Palaiseau, France

# Multiphase Averaging for Classical Systems

With Applications to Adiabatic Theorems

Translated from the French by H. S. Dumas

1988. XI, 360 pp. 60 figs. ISBN 3-540-96778-8

**Contents:** Introduction and Notation. – Ergodicity. – On Frequency Systems and First Results for Two Frequency Systems. – Two Frequency Systems; Neistadt's Results. – N Frequency Systems; Neistadt's Result Based on Anosov's Method. – N-Frequency Systems; Neistadt's Results Based on Kasuga's Method. – Hamiltonian Systems. – Adiabatic Theorems in One Dimension. – The Classical Adiabatic Theorems in Many Dimensions. – The Quantum Adiabatic Theorem. – Appendices. – Bibliography. – Bibliographical Notes. – Index.

Springer-Verlag
Berlin Heidelberg New York
London Paris Tokyo
Hong Kong

Springer